Plasmonics: Shrinking Metals, Expanding Possibilities

Jessy

Contents

1. Introduction and scope of the book

1.1. Introduction

By down-sizing the nobel metal from bulk to nano, it shows extraordinary properties such as nano-localized field enhancement, light confinement below the diffraction limit, highly efficient catalytic behavior, colour change as plasmonic pixels.[1–4] The reason behind all these striking properties is the oscillation of the conduction band electron inside the nanoparticle excited by external electromagnetic field. This oscillation is called localized surface plasmon resonance (LSPR) and the nano-object is called the oscillator (antenna).[5,6] Like any other antenna plasmonic nanoparticles do absorb in near field and scatter the light in far-field. Hence they have absorption and scattering cross-section.[7] The interaction between nano-object and photon can be tuned via altering the size, shape, material and surrounding.[8–10] As plasmonics is one of the fastest growing field in nanotechnology due to its fascinating and dynamic properties but plasmonic also comes with its inherent lossy nature.[11] it is always desirable to produce high optical quality (narrow full width at half maximum (FWHM)) systems with more plasmonic nature involved.[12,13] One possible solution to this problem is to utilize the superior quality plasmonic building blocks[14,15] namely with narrow size distribution, monodisperse and single crystalline in nature but colloids alone can't take full advantage of magnificent pool of applications.[16–18] It provides us with plethora of fascinating application such as plasmon heating, weak and strong coupling (plasmon lasing, Bose-Einstein condensation, lifetime enhancement), energy/ information transport and bio sensing, hot electron generation, are some from the wide range of application which intrigues the scientific community to dig deeper into it.[19–22] The excitement adds up with more futuristic application In the domain of optic, magnetic and other fundamental physical and chemical effects.[23–27] Unknowingly plasmonics was even used in 4th century as vivid colors of gold and silver colloid were used as colour pigment e.g., the famous Lycurgus Cup.[28]

1.1.1. Classical optics concepts:

Surface lattice resonance: The rudimental thought behind diffractive coupling of particle resonances was studied by several groups in the beginning of 1960s when Devoe presented the first theoretical model of an electric dipole for quasi-stationary aggregations of molecules.[29,30] Later Schatz and co-workers explained the physical essence of diffractively coupled effects and described conditions for the generation of ultranarrow resonances.[31,32] However the way to realize it experimentally was far from reality due to the high numerical aperture of focusing optics which leads to the low spatial coherence. There were other factors such as limited nanofabrication techniques as well which leads to more disorder into the system hence broader collective diffractive effects. The very first breakthrough in the experimental observation of ultranarrow collective plasmon resonances was obtained using optics that provided high spatial coherence to illuminate large arrays of Au nanoparticles.[33–35] Kravets et al. reported the first observation of ultranarrow plasmonic resonances (resonant widths down to 2–5 nm full width at half maximum, resonances with Q ~ 100) by illuminating, at a certain angle, a large array (30 × 60 μm) of glass substrate supported ~100 nm Au nanopillars. Auguié and Barnes[34] observed similar effects using normal incidence light transmitted through arrays of 50–120 nm Au nanorods, although the resonance features were generated only by placing the nanostructures in a uniform refractive index environment; very similar results were obtained at the same time by Chu et al.[35]

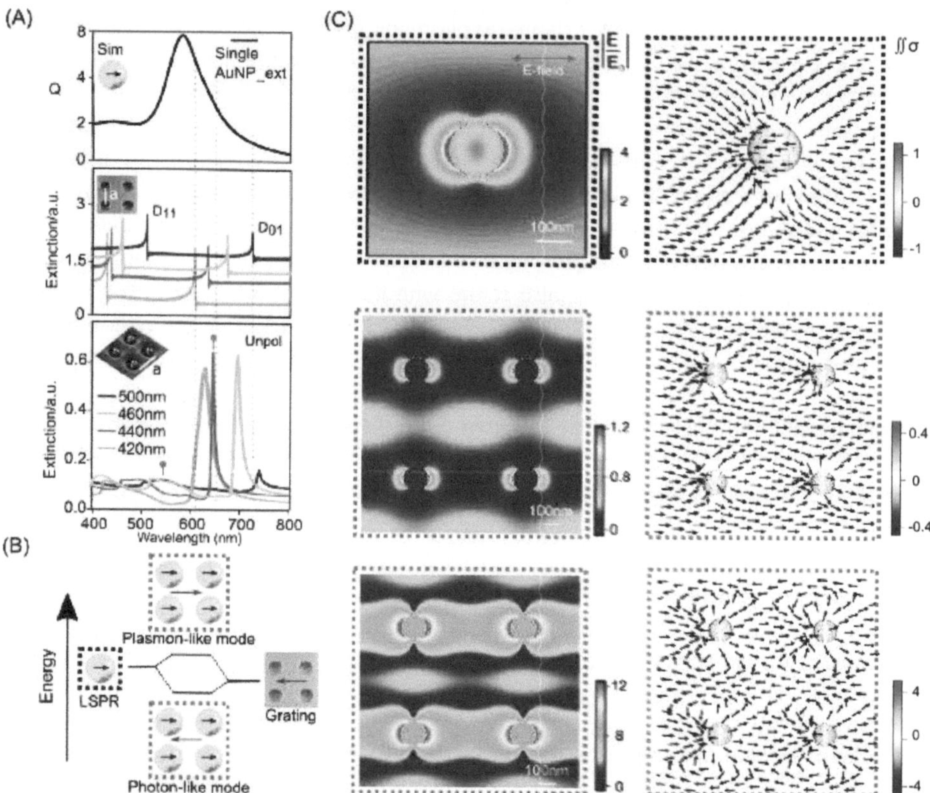

Figure 1: Optical properties of tunable surface lattice resonance: Variable and high optical quality lattice resonance by hybridization of plasmonic resonance and Bragg grating: (A) Simulated surface lattice resonance at various periodicities arise from coherent coupling and interference of LSPR hybridized by the first Bragg diffraction (D01). (B) Energy diagram showing the interaction between the LSPR and grating mode resulting in hybridized plasmon-like and photon-like mode. (C) Dashed frames showing selected normalized electric field images, surface charge images as well as electric field vector plots, which have been determined at highlighted wavelengths. Definition of abbreviations: effective extinction cross-section (Q). Integrated surface charge distribution ($\iint \sigma$) Adapted with permission.[4] [ ] Copyright 2019, *ACS Applied Materials & Interface*

The concept of variable surface lattice resonance originates from the coupling between plasmonic resonance of single particles and photonic resonance of a periodic Bragg grating

(Figure 1). At a distinct wavelength, the electron clouds of all particles in the lattice start oscillating in phase and the light scattered to the far-field interacts coherently. This results in effective trapping of light at the interface leading to a sharp peak in the extinction spectrum. We investigated the influence of lattice constant on the spectral response to optimize the SLR wavelength for a given particle system to maximize the quality factor. In order to get a sufficient spectral overlap, we choose gold nanoparticles with 80 nm in diameter and 2D square gratings with varying periodicities (a = 420, 440, 500 nm). Single gold nanoparticle dipole scattering mode is broad in terms of line width due to the strong radiative damping as depicted in (Figure 1A) by effective extinction cross-section (Q). This broad dipolar (bright) mode has a peak at 540 nm for 80 nm diameter gold nanoparticles in 'UV-PDMS' environment (considered as oscillator 1). In contrast, diffraction from a 2D grating becomes evanescent to radiative at the wavelengths

$$D(i,j) = \frac{n*a}{i^2+j^2} \tag{1}$$

An ordered array of metallic nanoparticles can be considered as a coupled oscillator system of previously discussed oscillators, namely single nanoparticle and grating. We have chosen the diffraction modes on the low energy shoulder of the LSPR resonance. In this energy range, the gold material shows low radiative damping. This coupled system supports two hybridized modes. The first mode is more plasmonic in nature or so-called plasmon-like, whereas the other hybridized mode has photonic features in it. These considerations are based on the fact that the electric field direction is flipped in the case of photon-like mode (SLR) due to the grating dominance as compared to the plasmon-like mode as shown in (Figure 1B). Simulated surface charge and inverted in plane e-field direction for plasmon-like and photon-like modes are shown in (Figure 1C red & blue dashed frames respectively). Moreover, due to coherent coupling, photon-like mode also allows nano-localized field enhancements, whose intensity exceeds that of the individual resonances. Figure 1C shows 3-fold e-field increase in case of photon like mode (blue dashed frame) as single particle LSPR (black dashed frame). Theoretical mode studies lead to the following conclusion: In order to obtain high quality hybridized modes, it is vital to maintain energetic stability which means that Bragg mode should be in the right shoulder of the single particle scattering shoulder.

Guided mode resonance: Figure 2 explains the working principle of a guided mode resonant (GMR) structure. The left column exhibits the geometrical representations of a GMR structure and its building blocks, the grating and the waveguide. The dimensions of these structures are matched to the electric field ($|E|$) profiles to the right (third and fourth) to correlate the spatial distribution within these geometries. Figure 2 (a) exhibits a sinusoidal grating of higher index $n_2 = n_{TiO2}$, backed by a lower index of ($n_1 = n_{air}$ = 1.00). The plot explains how the diffraction first order angle in transmission depends on wavelength for a fixed period of 520 nm and for normal incidence (θ_{inc} = 0°). This is quite straightforward and comes from the polarization independent grating equation

$$n_2 \sin\theta_D(m) = n_1\sin\theta_{inc} - m\frac{\lambda}{\Lambda} \tag{2}$$

In Figure 2(b), a waveguide of thickness 200 nm and index $n_2 = n_{TiO2}$ is associated with lower indexed superstrate $n_1 = (n_{pr} + n_{air})/2$ and substrate $n_3 = n_{glass}$. The indices n_{TiO2}, n_{pr} and n_{glass}

are wavelength dependent properties of titanium dioxide, , photoresist, and glass respectively, measured through spectroscopic ellipsometry and given in **publication VG2** Figure S1.

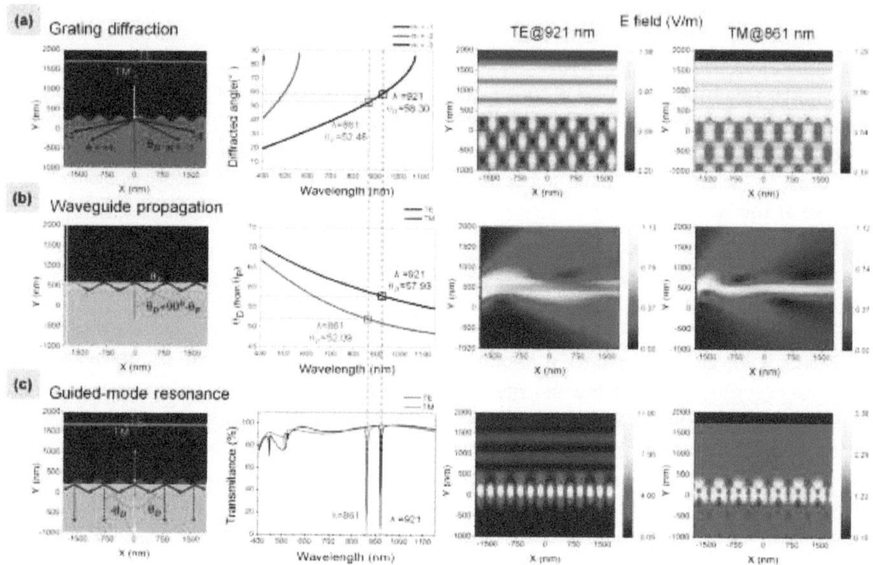

Figure 2: Physical interpretation of Guided-Mode Resonance (GMR) in terms of its building blocks: (a) Schematic of grating diffraction for a period of 520 nm. The plot shows diffracted angle vs wavelength for different transmission orders (b) Ray picture of waveguide propagation with propagating angle θ_P inside a slab waveguide. Plot shows θ_D (from θ_P) for TE and TM as a function of wavelength with a fixed waveguide thickness and constituting index. (c) GMR as grating plus waveguide structure. Ray picture shows diffraction leading to the propagation and coupling out. Destructive interference of these leaked out orders with zeroth order produce sharp transmission dips at 921and 861nm for the TE and TM case respectively. Electric field profiles ($|\boldsymbol{E}|$) at 921 and 861 nm are shown in (a), (b), and (c) showing diffraction of orders, waveguiding and strong resonance within waveguide correspondingly. Adapted with permission.[36] [ ] Copyright 2019, *ACS Applied Materials & Interface.*

The graph shows a plot of different propagation angle (θ_P) for different wavelengths expressed in terms of different diffracted angles ($\theta_D = 90° - \theta_P$) for the cases of transverse electric (TE) and transverse magnetic (TM) modes. These propagation angles are directly related to the wavelength dependent n_{eff} of the waveguide and can be calculated via numerical methods. Figure 2(c) shows the GMR structure interpreted as a grating together with a slab waveguide with and n_{pr}, n_{TiO2}, n_{glass} as grating, waveguide, and substrate indices. For a grating to decide the wavelength dependent diffraction angle (θ_D) that leads to propagation angle ($\theta_P = 90° - \theta_D$); only specific wavelengths depending on the polarization can produce these θ_P supported for the combined structure. As seen from the plots, the wavelength at 921 nm for TE mode only can give a diffraction angle of 58.3°as well as a propagating condition relating to 57.93° within this waveguide structure. The similar cases arise for the θ_D TM mode at 861 nm that satisfies both the case of diffraction and waveguiding for $\theta_D \sim 52°$. While propagating within the waveguide these selective wavelengths (TE and TM) interact with the gratings to diffract out and destructively interfere with the corresponding wavelength of the zeroth order transmission. These result in sharp GMR transmission dips as seen in Figure 2(c). The Electric field ($|E|$) profiles for the TE 921nm and TM 861 nm are shown in the right columns (third and fourth) corresponding to these individual cases and are described here. Figure 2(a) explains diffraction of a plane wave source upon encountering a high indexed sinusoidal gating. The modulus square of these orders at far-field can superpose to produce the well-known grating diffraction intensity pattern. Figure 2(b) depicts a cross-section of the angled-injection case for waveguiding of fundamental TE and TM modes. Figure 2(c) represents the case of guided-mode resonance within the waveguide with highly intensified electric fields as seen from the associated color bars. These modes can be interpreted as standing waves formed due to the interference of two counter-propagating diffracted waves referred to as the '+1' and '-1' orders.

1.1.2. Top down fabrication methods and their challenges:

There are several methods to produce nanostructures in this modern era. As soon as the new application aspect beyond information processing and storage in the areas of optics, biomedicine and material science will emerge there will be a demand for new fabrication technologies as well. Till date all methods are available to produce nanostructures. We can categories them in two parts: "top-down" and "bottom –up" methods. Mainly the top down approach consists of various lithography based methods which are serial and parallel in nature. They are able to produce nanoscale patterns typically in two-dimensions (2D) and on the length scale which are roughly 4 orders of magnitude larger than individual building blocks. There are several examples for such methods namely photolithography (with mask) , e-beam and focused ion beam lithography, laser interference lithography (without mask). On the other hand there are bottom up techniques which work on the interaction between molecules or colloidal particles to assemble discrete nanoscale structure in two and three dimensions.

Here first I would shortly explain the top down methods and further discuss the interferometric lithography in more detail. In microelectronics mostly two standard conventional lithography processes are used for patterning : scanning beam lithography and Photolithography. Scanning beam lithography is a serial process which is used to produce photomasks. The writing time with this procedure is dependent on pattern density and feature size. As an example, in order to write a dense array of sub-20 nm features over the area of 1 cm^2 requires near about 24 hours. There are three major classes of scanning beam lithography: (1) scanned laser beams with nearly 250 nm resolution (2) e-beam with sub 50 nm resolution and (3) focused ion beam (FIB) with sub-50-nm resolution. They can produce extremely high resolution nanostructures but all these processes are slow in nature relative to their counterpart (photolithography). Now in case of photolithography, this technique is widely used in the industries where a photomask produced via one of scanning beam methods is used to produce patterns with nearly 65 nm resolution. **As an example** A modern "step and-scan" photolithography system can pattern over one h**undred 300-mm** diameter wafers per hour with 65-nm resolution; it can also cost tens of milli**ons of dollars.**[37] In the contemporary world the size for semiconductor manufacturing process **is down to 5 nm** using deep ultavoilet (DUV) lithography and extreme UV (EUV) lithography te**chnique.**[38] **Here** in this work we mainly focus on interference lithography. Mostly all the pho**tolithography** methods demand for a photo mask which is typically patterned by e-beam **lithography.** This leads to the disadvantages in terms of higher cost requirement and **more time consumption.**

Table 1: Adapted with permission. [ ] Copyright **(2005), American** Chemical Society.

technique	current capabilities (2004)		
	minimum feature[a]	resolution	pattern
photolithography[1,b]	37 nm	90 nm	parallel generation of arbitrary patterns
scanning beam lithography[a,c]	5 nm	20 nm	serial writing of arbitrary patterns
molding, embossing, and printing[16,123,168,d]	~5 nm	30 nm	parallel formation of arbitrary patterns
scanning probe lithography[26,52]	<1 nm	1 nm	serial positioning of atoms in arbitrary patterns
edge lithography[59,e]	8 nm	16 nm	parallel generation of noncrossing features
self-assembly[354–357,f]	>1 nm	>1 nm	parallel assembly of regular, repeating structures

[a] Refers to the minimum demonstrated lateral dimension. [b] A resolution (pitch) of 45 nm is projected for 2010 using 157-nm light, soft X-rays, or optical "tricks" (e.g., immersion optics). [c] Obtained with a focused ion beam. Limited by photoresist sensitivity and beam intensity. [d] Limited by available masters and, ultimately, van der Waals interactions. [e] Potentially smaller sizes could be obtained using atomic layer deposition. [f] Self-assembly produces structures with critical feature sizes from 1 to 100 nm or larger.

These conventional methods such as e-beam are main**ly restricted to** planar fabrication and are incompatible with many problems such as th**ree-dimension** (3D) fabrication and scalability. These methods also use corrosive etchants, **high energy ra**diation and relatively high temperature. This is also true that these techniqu**es are highly en**ergy consumptive and researchers are looking forward to alternatives (unco**nventional)** techniques. Conventional methods are also limited towards selection of materia**ls when it comes** to patterning rather fragile materials and especially biological materials **these techniques** don't support that. Hence it is clear that regardless conventional or unco**nventional- any** technique which has capability to prototype nanoscale features rapidly and **inexpensively** will be acceptable in future. There is huge demand in the direction of real-to-**reel, parallel,** scalable, cost-effective and versatile methods.

1.1.3. **Template-assisted self-assembly**

It is very challenging to assemble these single plasmonic **building blocks in**to a defined manner. Even after the assembly, the emerged modes such as **gap mode, particle** chains longitudinal mode (L mode & L* mode) are broad and lossy in nature. **Second possible** solution is to feature a high-quality factor (Q-factor) photonic mode in combi**nation with plas**monic mode. In other words, the assembly should fulfill a purpose in accordan**ce with the requi**rement of photonics to be excited at the same time with plasmonics. The **interaction between** photonic and plasmonic leads to new hybrid modes where one can t**ake the full adva**ntage of plasmonic by keeping the losses as low as possible into the system. **Fano resonance** is one of the best examples for such an oscillator. There are several tech**niques available t**o produce plasmonic metamaterial with high resolution. Many of these techniques arise from the combination of photolithography, etching, electron-beam (e-beam) deposition and lift off. Since all these methods lack the fundamental requirement of having single crystalline building blocks, and monodispersity. Here in this book we shift our focus towards directed self-assembly which could serve the purpose of being scalable, highly parallel and rapid.

There is a need to define self-assembly itself and make a distinction between SA and directed SA. Self-assembly is a phenomenon where spontaneous organization of the nanoparticles into complex structures occurs. This method is facile, parallel and scalable as compared to other e-beam related fabrication methods. However, self-assembly is very often restricted to environmental conditions and sensitive to particle type, particle coating and interactions between them. This raises the question about robustness of the system. Later directed self-assembly comes into focus which as a part of self-assembly leads to the solution of some of the basic problems. There could be several methods to direct or drive the particle to have precise placement at nano scale with high yield and robustness. As colloidal lithography[39], a collection of methods based on the use of colloidal structures for pattern replication, is a relatively new research field, the goal of this review is to explore the already available techniques for the creation of self-assembled colloidal crystals. Colloidal particles are an intensively studied material as an option for the building blocks required for self-assembly[40–42], ranging in size from a few nanometers up to a single micron, their small size renders them subject to Brownian motion.[43] However, this also makes them readily available to be suspended in a variety of liquids, forming so-called colloidal suspensions.[44] Their most attractive feature is their tendency to form crystalline organizations, which allows them to be used as intermediate building blocks for larger and more complex structures.[45]

Colloidal particles: Several techniques and protocols have been developed for the synthesis of colloidal particles. Depending on the used material (e.g. gold, silica, polystyrene), colloidal particles can be synthesized by suspension-, emulsion-, dispersion-, and precipitation-based polymerization[46] methods. In general, factors that affect the size and polydispersity[47] of particles are pH value, the concentration of catalyst, the composition of reagents, the type of solvents, and the reaction temperature.

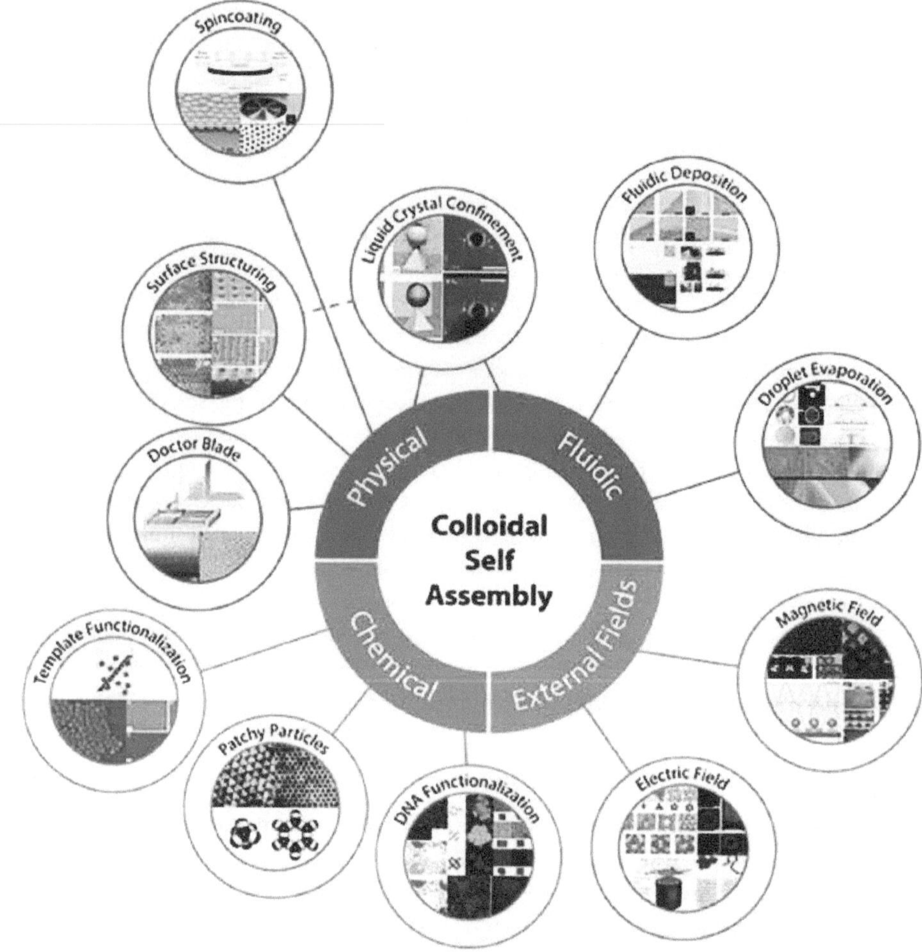

Figure 3: Scheme of colloidal self-assembly techniques. Each of the 4 domains (physical, fluidics...) indicate the origin of the main force which drives the process. Adapted with permission.[48] [ ] Copyright **2018** *Advances in Colloid and Interface Science.*

Several reviews offer a complete overview of colloidal particle synthesis and characterization.[49–53]

Importance of surface functionality: It is important to mention that functionalizing the surface of a colloidal particle may also alter its surface chemistry.[54] The main usage for colloidal surface functionalizing is to prevent agglomeration, with common functional groups being hydroxyl ($-OH$) and carboxyl ($-COOH$) chemical moieties. Particles can also be encapsulated by a metal or oxide shell to allow for further functionalization.[55]

Interactions between colloidal particles: The behavior between colloidal particles is governed by a large number of interactions, and the colloidal crystal formation and stability result from a balance between the different forces. A brief overview of the most important interactions is given here, both between particles surrounding environment.[56–59]

van der Waals interaction (vdW)[60]: vdW interactions result from the interaction between the individual constituent of the particles. It is possible to exploit this energy to drive self-assembly, however this works most effectively only for small, nanosized, particles. vdW interactions rely on short range forces that are overcome by other forces for surface separations exceeding 10 nm. As interaction distances increase to larger ranges, the influence of vdW interactions falls off rapidly.

Electrostatic interaction[61,62]: Electrostatic interaction occurs due to particle surface charges and can be either attractive or repulsive. When particles are suspended in a liquid, the electric double layer effect comes into play, creating an electric field and therefore yielding a slight repulsion between two or more particles. The decay of this effect's influence is usually expressed in terms of the Debye screening length or screening effect; it depends on the electrolyte concentration, pH and particle concentration. It is important to mention that electro-kinetic phenomena due to the movement of charged particles can be very influential during colloidal crystal formation.

Steric effect[63]: Colloidal particles in highly concentrated media are subject to steric effects[64], where the presence of other particles in the liquid physically blocks further ordering or material deformation. One key steric effect is depletion interaction, which occurs when particles suspended in a solution approach each other and a solute molecule no longer fits in between them. The particles are then pushed together, as the surrounding molecules of the liquid exert an osmotic pressure[65]. This effect is especially prevalent in macromolecular solutions, such as liquid crystal[66] and polymers.

Capillary forces[67] Attractive capillary forces originate due to the formation of liquid bridges[68], typically observed in meniscus effects, forming between particles due to surface tension. Their behavior can be modelled by Laplace Pressure. A large number of additional forces can also play a role in colloidal self-assembly, including structural forces, fluctuation wave forces, bridging interaction and solvation forces[69].[56] The self-assembly methods are organized based on the type of dominant force that is mainly responsible for the particle assembly

There are four main categories as shown in (Figure 3):

- Physical: self-assembly process dominated by shear forces, adhesion and surface structuring;
- Fluidic: self-assembly process dominated by capillary forces, evaporation, surface tension;
- External Fields: self-assembly process dominated by electric and magnetic fields;

- Chemical: self-assembly process dominated by chemical interaction, changing the surface charge or creating binding sites.

This external drive is the cause for complex assemblies and fascinating effects. This method is cheap, scalable, faster and programmable as compare to pick and place techniques *e.g.* AFM or optical tweezers. Appealing methods for achieving such plasmonic structures are interface-mediated and capillary-assisted assembly. The interface-mediated assembly method is a powerful tool for creating hexagonally ordered arrays. Additionally, the stacking of these arrays further extends the range of accessible structures to more advanced ones like honeycomb or Moiré lattices.

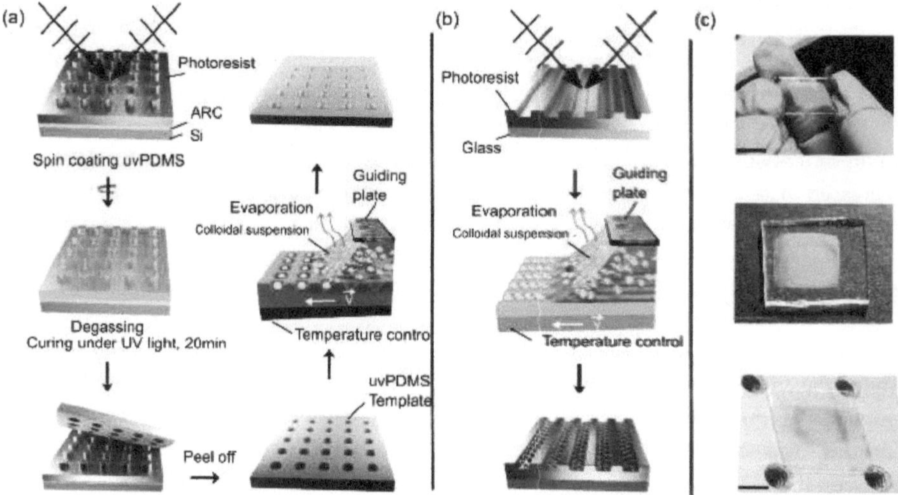

Figure 4: Capillarity-assisted particle assembly with soft-lithography templates. a) Scheme representing the fabrication of a stretchable 2D lattice using a combination of laser interference lithography with double exposure, soft molding with PDMS, and directed self-assembly of plasmonic nanoparticles. b) Scheme represents the direct assembly of colloidal nanoparticle (AuNP) inside nanochannels. C) Template showing the 2D nanopillars, replicated 2D nanoholes on uvPDMS and assembled 1D nanochannels on photoresist grating on glass respectively. Scale bar is 1 cm.

Furthermore, interface-mediated assembly can be used to create masks for metal deposition. However, while these methods provide a simple method for large-scale assemblies, they are also limited by their underlying reliance on hexagonally ordered particles. Assemblies with a broader range of symmetries, more pronounced anisotropy, or local variation of the packing can be achieved via template-assisted self-assembly (TASA).[70] Colloidal self-assembly is a thermodynamically driven process resulting from the interplay of different attractive interactions (e.g., van der Waals forces, Coulomb attraction, and depletion forces) and repulsive interactions (e.g., Coulomb and steric repulsion). Local energetic minima can be introduced by applying templates featuring attractive electrostatic interactions or support–

capillary interactions that direct the particles during the drying process. To reach these minima, external driving forces can be applied via confinement, convective flow, electric/magnetic fields, or templating molecules (DNA).[48] This technique is known as directed self-assembly and leads to particle deposition in a desired fashion.[45,71,72] self-assembly methods for the creation of colloidal crystals, organized by the type of forces governing the self-assembly process: fluidic, physical, external fields, and chemical. The main focus lies on the use of spherical particles, which are favorable due to their high commercial availability and ease of synthesis. However, also shape-anisotropic particle self-assembly will be introduced, since it has recently been gaining research momentum, offering a greater flexibility in terms of patterning.

The mechanism of particle assembly through drying of particle dispersions on flat substrates has been studied intensely by the group of Nagayama[73,74] It was found that the increased solvent evaporation close to the three-phase contact line induces a convective flow of particles toward the contact line. This phenomenon is often referred to as the "coffee ring effect" and can be utilized for particle assembly. The assembly processes based on convective flow have been foremost interpreted by Wolf and co-workers[75,76] who refined the common dip-coating process. As shown in Figure 4a (center right), a droplet containing dispersed particles is confined between the template and a glass slide. Nanoparticles accumulate at the meniscus as a result of the convective flow. By withdrawing the template, the meniscus is moved across the substrate leading to particle deposition in the grooves of the template. The parameters to control in these experiments are predominantly the movement speed, the substrate wetting (by surfactants), and the dew point. Two cases are distinguished for the assembly: the convective and the capillary regime. The convective assembly occurs at very low contact angles (<20°) in which the particles are predominantly confined by a remaining thin solvent layer.[75] This film forces them into the recesses during the evaporation, similar to the mechanism during spin-coating. Conversely, the capillary regime describes the assembly at comparably high contact angles (>20°). The meniscus is not continuously moving but pinning at the topographical features. As a result, capillary forces drag particles into the recesses where they stay confined after unpinning. By rational design of the features, the orientation of anisotropic particles can be predefined. For instance, an energetic minimum can be reached by nanorods aligning parallel in rectangular grooves.[77] By directing particles into accurate positions, this tightly controlled method also enables the deposition of single particles in an array.[78] As a big advantage of this assembly method, the assembly area is only limited by template size and reservoir of colloidal particle solution. It was already demonstrated that the latter limitation can be overcome by a continuous microfluidic feed of suspension.

The technique is applicable to various types of topographically structured templates, including elastomeric substrates.[79] Thus, it is an excellent foundation for the preparation of mechanotunable colloidal assemblies. Suitable templates for colloidal assembly can be prepared by various methods of which we want to highlight laser interference lithography (LIL).[80] Figure 4a depicts the use of LIL in combination with CAPA to produce a plasmonic lattice. Gupta et al. employed LIL to produce large-scale structures with nanometer precision at low cost as compared to e-beam lithography. Lloyd's mirror technique was used where a single coherent laser beam is converted into two coherent sources. Interference of the two beams occurred on a photosensitive material. By a second exposure with 90° rotation, a nanopillar structure was formed after curing (see Figure 4a). This master was replicated into a nanohole structure using PDMS. With this template, 80 nm diameter gold spheres were assembled in a square lattice of single nanoparticles as displayed by the black frames. Further in figure 4b direct assembly of photoresist has been done to fabricate nanochannels.

1.1.4. Functional colloidal surface assemblies:

Recently, a few studies have emerged reporting tunable metasurfaces for light modulation working at optical frequencies.[27,81–83] Ability to being Flexible, tunable, stretchable, bendable becomes some of the crucial parameters in device fabrication. PDMS emerges as a suitable candidate for tunable nanophotonic devices[84] because mechanical deformation of photonic nanostructures fabricated on flexible substrates allows for simple, quick, and reproducible tuning of photonic devices.[4,85–88] Active plasmonics becomes one of the most promising area of research in last decades.[89] The dynamic control over plasmonics have been shown in several occasions by thermal[90], acoustic or mechanical stimuli. This book is mainly devoted to mechanical stimulus as a trigger for so-called plasmo-mechanics. In this field the tunability of the plasmonic effect is usually achieved by exploiting the elastic properties of a flexible substrate. The sensitivity and accuracy of optical responses toward mechanical deformation (and other triggers) can be modified favorably by a careful arrangement of the plasmonic particles. The afore-mentioned approaches share the following characteristics: On the one hand, the response behaves nonlinearly, with high sensitivity at low strain and decreasing feedback at higher strains. On the other hand, the plasmonic resonance (especially in the coupled state) is comparably broad (mediocre quality factor). A sharper peak with a high quality factor would greatly improve the readout and detectability of minor changes. Both conditions can be overcome by utilizing SLRs instead of LSPRs for creating optical feedback. The combination of elastomeric substrates and lattice resonances has been foremost studied by the group of Odom,[85,86,91] who investigated the behavior of different lattice geometries prepared. Aside from this lithography-based work, colloidal self-assembly enables the preparation of plasmonic lattices on elastomers like PDMS. Simple designs cost-effectively fabricated via bottom-up techniques even stand the comparison with e-beam structures for nano-plasmonic devices. Additional tunability of the strong optical effects by external stimuli facilitates broad-band applications and opens the door for sensing applications. For an example as shown in figure 5A an actively tunable collective localized surface plasmon was achieved by responsive hydrogel membrane.[92] in figure 5B Habib et. al developed a method to selectively and conformally electro polymerize a metallic nanoantenna with PEDOT:PSS loads. By performing electro-optic measurements from high mode volume, they demonstrated enhanced field sensitivities with electrochromic loading and establish high signal-to-shot noise ratio (SSNR) measurement capability from single electro-plasmonic nanoprobes.[93]

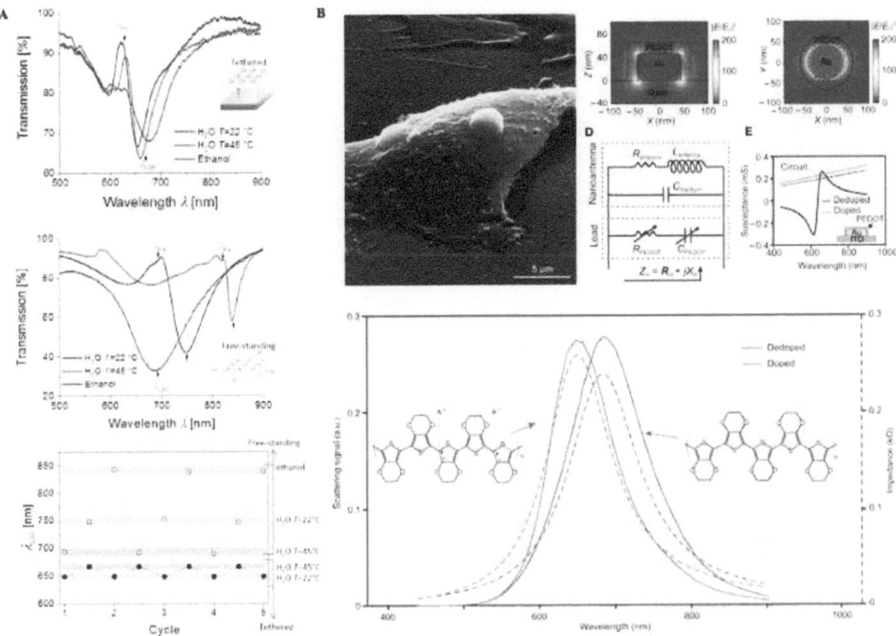

Figure 5: Stimuli triggered electro-plasmonic nanoantenna. (A) Transmission spectra measured for tethered structure (A, prepared with Λ = 460 nm, D = 110 nm) and free-standing structure (B, prepared with Λ = 360 nm, D = 140 nm) in contact with water at T = 22 °C (blue curve), T = 45 °C (red curve), and in contact with ethanol (green curve) at room temperature for the normally incident beam θ = 0°. Measured changes in the wavelength λ_{LSP} upon a series of five cycles of swelling and collapsing in H2O and in ethanol for tethered (bottom) and free-standing (top). Adapted with permission.[92] [ ] Copyright **2019** *Adv. Optical Mater.* (B) SEM image of cardiomyocyte cells cultured on an array of electro-plasmonic nanoantennas. Considerable size difference between loaded nanoantennas (height, 45 nm; diameter, 90 nm) and electrogenic cells is shown. A total of 2.25 million electro-plasmonic nanoantennas are incorporated on a transparent substrate with nanometer spatial resolution, allowing measurement of electric-field dynamics from diffraction-limited spots over a large surface area. Side view of near-field enhancement $|E/E_0|^2$ along the pristine nanoantenna at 678.8 nm. FDTD simulations show that plasmonic excitations lead to strong confinement of the light within the 20-nm-thick electrochromic layer. Top view of the near-field enhancement $|E/E_0|^2$ profile along the center of the pristine nanoantenna at 678.8 nm. Equivalent nanocircuit model of the electro-plasmonic nanoantenna. Electrochromic doping is incorporated through tunable resistor and capacitor elements. Susceptances of the gold nanoantenna and the PEDOT: PSS load for doped (red) and dedoped states (blue) are shown. Intersections (indicated by the circles) correspond to the open-circuit condition, the plasmonic resonance. For the doped (dedoped) electrochromic load, the resonance condition occurs at the shorter (longer) wavelength intersection due to the diminished resistance (losses) of the electrochromic load. Far-field response of the electro-plasmonic nanoantenna to the doping state of electrochromic load. Electrochromic switching of the load from the doped (red curve) to the dedoped (blue curve) state leads to red shifting of the plasmonic resonance. FDTD simulations (solid curves) and

lumped nanocircuit model (dashed curves) are compared. The inset depicts the chemical structure of PEDOT for the doped (left) and dedoped (right) state. A⁻ represents the counterions. Adapted with permission.[93] [ ] Copyright **2019** *Sci. Adv.*

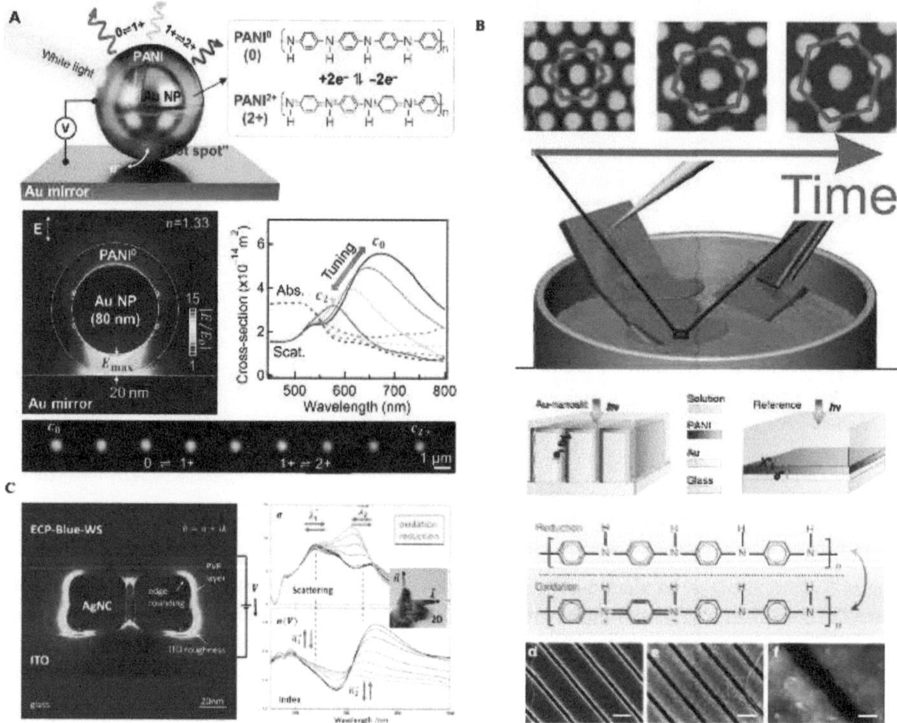

Figure 6: Electrochromic plasmonics. (A) Schematic of an *e*NPoM, which changes color as a function of redox state of the thin (0 to 20 nm) PANI shell surrounding each Au NP on Au mirror substrate. Right: Redox reaction of PANI in the gap (PANI⁰, fully reduced; PANI¹⁺, half oxidized; PANI²⁺, fully oxidized). Optical near-field enhancement of the *e*NPoM for reduced state of PANI shell (PANI⁰) showing hot spot in gap and its corresponding optical scattering (solid lines) and absorption spectra (dashed lines) for different redox states of the PANI shell (red to green: PANI⁰ to PANI²⁺), from numerical simulations. Experimental dark-field (DF) scattering images of a single *e*NPoM nanopixel for different redox states of the PANI shell (left to right: PANI⁰ to PANI²⁺). Adapted with permission.[82] [ ] Copyright **2019** Sci. Adv. (B) Macroscopically sized, hexagonal monolayers with exceptionally high degrees of order are fabricated in one step. Time controls the interparticle spacing and no further processing is required. Adapted with permission.[94] [ ] Copyright **2015** *Adv. Mater.* (C) Electrically tunable plasmonic behavior of nanocube-polymer nanomaterials induced by a redox-active electrochromic polymer. Adapted with permission.[95] [ ] Copyright **2014** *ACS Nano.* (D) Schematic diagram of a plasmonic electrochromic electrode incorporating Au-nanoslit array and reference planar electrochromic electrode. The pitch of the Au-nanoslit array is 500 nm. The depth and width of the slit is 60 and 250 nm, respectively. Chemical structures of PANI in the reduced and oxidized form. SEM images of the fabricated Au-nanoslit electrode (d) before and (e) after deposition of a PANI to a thickness $d \approx 15$ nm. (f) Magnified

SEM image from e. Scale bars, 300 nm (d,e). Scale bar, 100 nm (f). Adapted with permission.[96] [ ] Copyright **2016** NATURE COMMUNICATIONS.

With these examples it is evident that in terms of triggers; active plasmonics is not limited to just mechanics. There are a number of functional polymers i.e. smart polymer, shape memory polymer, enzyme-responsive polymer, light responsive polymer[97], pH-responsive polymer[98], thermo-responsive polymer[99] etc. In figure 6A authors have reported electrochromic nanoparticle-on-mirror constructs (eNPoMs) formed from gold nanoparticles (Au NPs) encapsulated in a conductive polymer shell [here polyaniline (PANI)]. Our scheme works by switching the charge state of the entire PANI shell, thus rapidly shifting the resonant scattering color of the eNPoM across >100-nm wavelength ranges. This active nanopixel only requires ~0.2 fJ of energy for each 1-nm shift in wavelength and can achieve commercial video rates (20). We show that centimeter-scale eNPoM metasurfaces assembled into disordered patterns by a scalable directed self-assembly scheme show vivid uniform color dynamics, not possible with any existing plasmonic color system.[82] However volk et al. has showed time-controlled colloidal superstructure where collective coupling was achieved in particle monolayers figure 6B.[94] Similarly a silver nanocube Langmuir-Blodgett (LB) monolayer that is coated with an electrochromic polymer enables reversible refractive index variation under variable electrical potential, which causes reversible shifts in plasmon resonance peak positions of silver nanocubes (figure 6C). Further a high-contrast and fast electrochromic switching enabled by plasmonics (figure 6d,e,f).[96]

Moreover, we have shown that colloidal assembly strategies are well suited for the preparation of periodic plasmonic particle arrays that can support surface lattice resonances as the consequence of plasmonic/diffractive coupling. The progress in the field is driven by advances in modern particle synthesis: Core/shell particles with plasmonic cores and soft polymeric shells are ideal colloidal building blocks for various assembly strategies where the shell can be used as a dielectric spacer and thus defines the lattice periodicity. Furthermore, the role of the polymeric shell can go beyond structural control by integration of polymers with responsiveness towards environmental parameters such as solvent, pH, light or temperature or complementary functionality such as by the integration of fluorescent or conductive polymers. These core/shell particles can be assembled into low-defect 2D lattices on fluid interfaces and subsequently transferred onto solid supports. Alternatively, the use of patterned substrates offers possibilities to direct colloidal assembly and thus further increase the pattern complexity. Here we have especially emphasized approaches that are scalable, such as controlled wrinkling or interference lithography. Finally, multiple transfer or printing steps allow for the assembly of complex multi-layered architectures. While classical lithography-based approaches for the fabrication of such plasmonic arrays are limited in terms of the total array dimensions, colloid-based fabrication schemes can overcome these limitations. At the same time, colloidal assembly is more energy-efficient and environmentally benign, as it is typically relying on synthesis in aqueous media and assembly rather than etching steps or processing using organic solvents. Finally, we aim towards the scope of the book to have scalable colloidal assembly concepts for responsive and functional

1.2. Scope of the book:

This book combines current concepts in colloidal self-assembly with current concepts from photonics in order to obtain large-area nanostructures with emergent spectroscopic properties. The scope of this book is to understand and investigate the coupled plasmonic systems. In this context, the basic ideas originated from optics such as guided mode resonance (GMR), Bragg diffraction are combined together with plasmonic colloidal building blocks such as localized surface plasmon resonance (LSPR) and particle chain modes. This admixture leads to the high-quality coupled systems namely surface lattice resonance (SLR) and hybrid GMR. In short, first we target to understand the basic plasmonic systems ranging from local (single nanoparticle) to colloid coupling (hot spot / plasmon hybridization) and collective colloid coupling (particle chains / plasmonic waveguiding). An interesting aspect of colloidal assembly methods is the intrinsic compatibility with stretchable or even mechano-tunable devices. Particle assemblies can be readily transferred onto stretchable substrates such as elastomers and the modular/ particle-based architecture is ideal for reversible mechanical deformation. Deformability is especially interesting for controlling chiroptical structures and surface lattice resonances, as the periodicities and symmetries in the arrays can be altered by the deformation. We discuss the perspectives of this mechano-tunability for strain sensing and dynamic tuning of the resonance wavelength.

So far, Optical concepts have already been shown by top-down lithography. There are other colloidal approaches, such as the floating method. These are not scalable/stretchable due to domain formation. That is why this book aims to fabricate plasmonic-photonic systems and realize the effects in a cost-effective, scalable and robust manner. In this context, we optimize for the amalgam of top-down and bottom up approach. We introduced a mask less, rather low cost and quick method of laser interference lithography (LIL) and soft molding to produce large-scale flexible one- and two-dimension templates. Further in this context, colloidal self-assembly offers an advantage over conventional top-down methods in terms of providing low cost and compatibility with roll to roll mass production. By means of self-assembly, these colloids can be selectively placed into their potential energetic minimum in given structures. Electromagnetic simulation methods allow the spectroscopic properties to be determined and suitable structures to be calculated. Such an approach of rational design, large-scale implementation and characterization to obtain surfaces with emergent properties is also called the colloidal surface concept **(see chapter 2.1)**.

In these colloidal surfaces strong light matter interactions have been shown under TM polarization but here we prove that similar effects can be achieved under TE excitation **(see chapter 2.2)**. The purview of this book and the application aspect of the hybrid modes are executed in understanding of hot-electron generation. We show that the interplay between confined photons on a semiconductor waveguide and localized plasmon resonances on nanoparticle chains modifies the efficiency of the photo-induced charge-transfer rate of plasmonic derived (hot) electrons into accepting states in mechanotunable systems. We aim to introduce flexibility in template fabrication and focus to introduce mechanically induced tunability in hybrid-optical systems. This has a huge potential in terms of integration with flexible electronics. We focus on strain induce SLR tuning where non-degenerate SLR modes are produced and toggle in between with the help of polarization. Further plasmonic chirality is triggered in bi-layer plasmonic chain structure and compression induce real time chiro-optic modulation is achieved. **(see chapter 2.4).** Further gold is deposited inside the molds using a seeded-growth procedure. By carefully exploring the growth parameters we show that gold nanostructures with aspect ratios of up to 7 can be grown from single seeds.[100] Further the strong coupling has been achieved in 1D colloidal out of plane lattice resonance and film coupled cavity mode. Additionally, **real** time tunable plasmonic chirality has been proved in natural molecules in 1D nanoparticle assembly.

2. Results and Discussion

2.1. Mechanotunable Surface Lattice Resonances in the Visible Optical Range by Soft Lithography Templates and Directed Self-Assembly

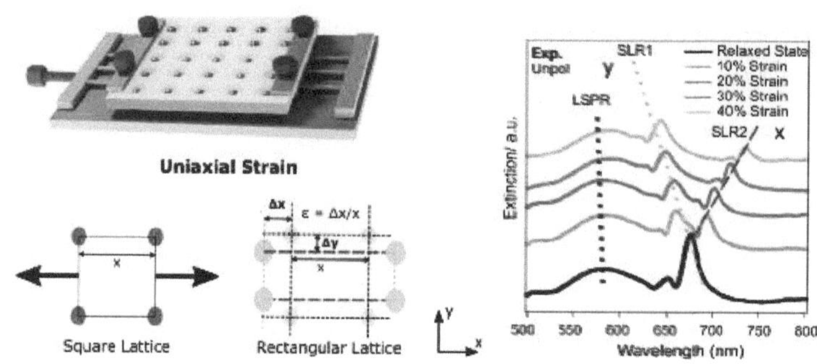

This section is based on the peer reviewed journal article. Adapted under the terms of ACS Author Choice license.[4] [ ] Copyright 2019, *ACS Applied Materials & Interface.*

By **Vaibhav Gupta**, *Patrick T. Probst, Fabian R. Goßler*, **Anja Maria** *Steiner, Jonas Schubert, Yannic Brasse, Tobias A. F. König,* * *and Andreas Fery**

V. Gupta, Fabian R. Goßler, Anja Maria Steiner, Patrick T. Probst, Jonas Schubert, Yannic Brasse, Dr. Tobias A. F. König, Prof. Andreas Fery
Institute for Physical Chemistry and Polymer Physics, Leibniz-Institut für Polymerforschung Dresden e.V., Hohe Str. 6, 01069 Dresden, Germany.
Patrick T. Probst, Jonas Schubert, Yannic Brasse, Dr. Tobias A. F. König, Prof. Andreas Fery
Cluster of Excellence Center for Advancing Electronics Dresden (cfaed) and § Physical Chemistry of Polymeric Materials, Technische Universitat Dresden, 01069 Dresden, Germany.

∥ Equal contribution: V.G., P.T.P.

* Corresponding author

Author contribution statement

VG and PP contributed equally to the preparation of **the manuscript. VG** and FG did the template fabrication including the optical set-up for laser **interference** lithography and soft molding. **VG** and PP performed the self-assembly experiments. **VG and** YB carried out optical spectroscopy and stretching experiments. **VG** and TK performed the FDTD simulations. AS and JS contributed with the gold nanoparticle synthesis and different coating protocols. TK, AF, **VG**, PP supported the manuscript writing process and were involved in the scientific discussions. TK and AF contributed in developing the concept and writing of the manuscript.

Introduction

Plasmonic resonances of metallic nanoparticles are of broad interest because of their strong interaction with light and their energy confinement at sub-wavelength scales. Light interaction induces collective oscillation of free electrons commonly known as localized surface plasmon resonance (LSPR). Spectral properties of such modes can be engineered by varying the nanoparticle composition, size or shape, and the surrounding refractive index.[15] For photonic and optical applications, it is desirable to achieve high optical quality and wide optical tunability simultaneously. The quality factor is defined by the ratio of the resonance wavelength to the resonance width. One effective approach toward obtaining high optical quality is the interference between a plasmon mode and a Bragg grating mode, also known as symmetry breaking or Fano resonance.[12,101] The arising surface lattice resonance (SLR) yields coherence to the system by confining the high electromagnetic field in the lattice plane.[102] Whereas typical LSPR exhibits a full width at half-maximum (fwhm) of >80 nm, for SLR the spectral width is drastically reduced, down to 1–2 nm.[103]

In order to steer this lattice plasmonic resonance to the visible wavelength (λ), an energetic overlap between the Bragg mode and the particle plasmon resonance is necessary. In the first order and at normal incidence, the mode occurs at λn (n as environmental refractive index).[104] To cover the entire visible area, the size of the particles and the lattice periodicity must be adjusted accordingly.[10] Such SLR response in the visible wavelength range was demonstrated for a lattice of aluminum rectangles fabricated by electron beam lithography.[84] In those examples, lithography methods were employed, which require masks and etching processes. Thus, fabrication becomes costly and highly energy-consumptive. Colloidal self-assembly can be a more cost-efficient and ecofriendly alternative, and renders SLR on cm^2 areas easily accessible. For example, non-close hexagonally packed particle arrays supporting SLR can be readily generated by floating plasmonic core–shell nanoparticles at the water–air interface.[105] In this system, the temperature-induced change in the refractive index of the hydrogel shell additionally allows a reversible 50 nm shift of the hybridized mode with a line width of 50 nm. Recently, mechanical deformation of plasmonic arrays sparked research interest as an alternative stimulus for spectral tuning.[103] Odom and co-workers showed that optimum conditions for lasing can be achieved by mechanical tuning.[86] As well, we showed in earlier work that colloidal linear particle assembly can be split into plasmonic oligomers, resulting in strain sensitivity and potential applications in sensing.[106] The colloidal approach is particularly well suited for creating mechanically tunable assemblies as particles can be directly transferred to the target substrate or embedded in elastomers. For precise strain-induced tuning of optical properties, it is mandatory to address specific lattice orientations consistently over the whole substrate. Therefore, defect-free templates molded from electron beam masters are commonly used in capillarity-assisted particle assembly of two-dimensional (2D) lattices.[76,107] Porous anodic alumina[108] or laser interference lithography (LIL)[36] can be a better scalable alternative. In contrast, hexagonal assemblies produced template-free at the water–air or liquid–liquid interface suffer from differently oriented domains. The variation in strain-induced lattice deformation from domain to domain would lead to a broad averaged SLR peak.

In this work, we combine the top-down method soft lithography and the bottom-up method template-assisted self-assembly (TASA) as facile and scalable fabrication methods to generate a mechanically tunable SLR. Flexible nature of the metasurface can overcome the optical cavity design limitation by creating real-time tunable high-quality mode. To design the collective optical properties for the visible wavelength regime, we use electromagnetic finite element simulations. After realizing the square lattice of gold nanospheres assembled inside an elastomeric template, the sharp spectral features, modulated by mechanical stresses, are quantified using conventional UV–vis spectroscopy. Both our theoretical and experimental results stress the importance of particle size polydispersity for the quality of the collective SLR mode. Our approach of flexible and tunable high-quality plasmonic lattice paves the way for cost-effective next-generation optical and photonic devices.

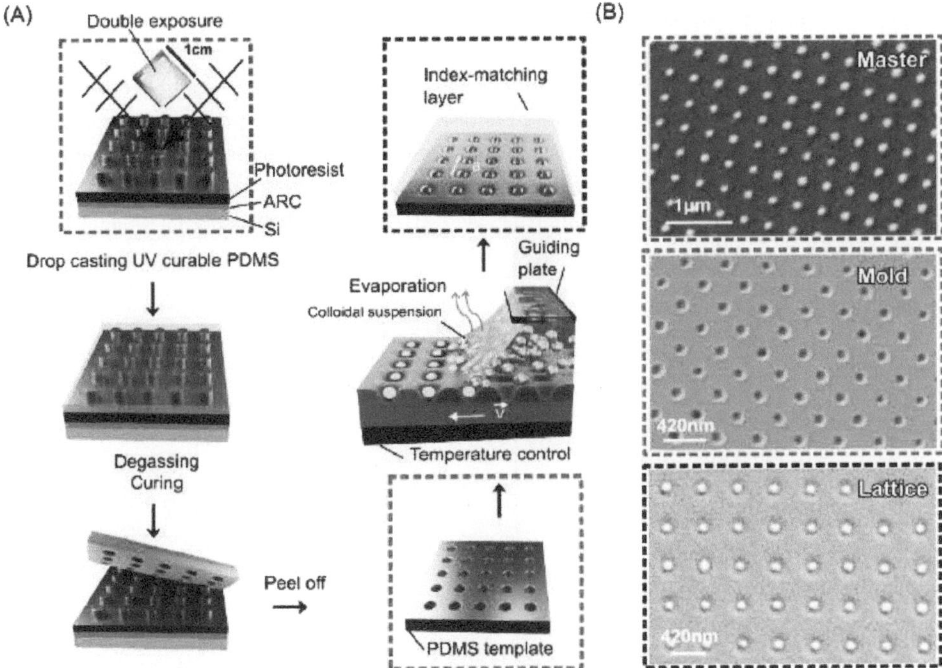

Figure 7. Soft lithography template fabrication for directed self-assembly. (A) Scheme of a stretchable 2D plasmonic lattice under uniaxial strain. Corresponding TEM images of two particle systems with broad and narrow particle size distribution. (B) Scheme represents the fabrication of a stretchable 2D lattice using a combination of LIL, soft molding, and directed self-assembly of plasmonic nanoparticles. ARC: anti-reflective coating. (C) SEM images of the fabricated nanopillars master, nanoholes array, and 2D plasmonic lattice (narrow size distribution). Dashed frames in pictorial representation correspond to the SEM image. Adapted with permission.[4] [ ] Copyright 2019, *ACS Applied Materials & Interface*

2.1.1. Fabrication of flexible 2D plasmonic lattice

We followed a soft lithography fabrication approach to create a square lattice of well-defined periodicity over cm^2 areas (Figure 7B). This method starts with a double LIL exposure to fabricate the nanopillar master on a photoresist-coated silicon wafer. Negative replication by a soft lithography molding step produces a nanohole structure (nanohole array from UV-curable PDMS). This flexible template was used for capillarity assisted assembly to align the colloidal plasmonic nanoparticles in a 2D square lattice. Briefly, a droplet of nanoparticle suspension was dragged across the topographical template. The hole dimensions were chosen (120 nm diameter and ~70) such that only one particle fits into it. At the receding meniscus, the particles concentrate facilitated by evaporation-driven convection. This local accumulation reduces the Brownian motion of nanoparticles and thereby increases the probability of particles being deposited inside the features. Net repulsive particle– substrate interaction and good colloidal stabilization of particles provided by surface modification prevent random sequential adsorption to the substrate and irreversible aggregation, respectively. Precise tuning of contact angle and surface tension balances the components of the capillary force at the meniscus acting parallel and normal to the substrate to drag the gold nanoparticles along and only allow deposition where they feel a sidewall of the topographical features counteracting this motion. As a big advantage of this assembly method, the assembly area is only limited by template size and reservoir of colloidal particle solution. It was already demonstrated that the latter limitation can be overcome by a continuous microfluidic feed of suspension.

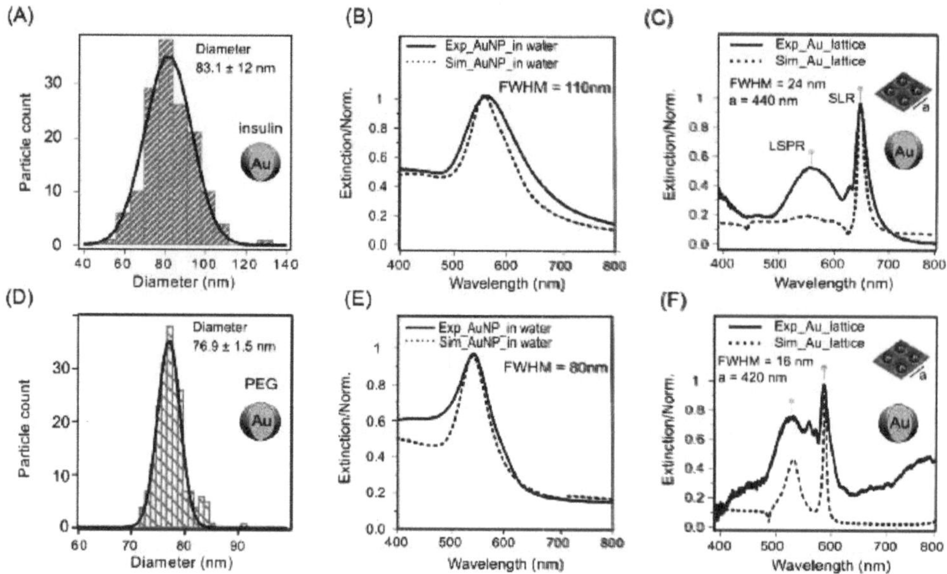

Figure 8. From various colloidal particle systems to SLR. (A, D) Size distribution of broadly/narrowly distributed gold nanoparticles coated with insulin and PEG, respectively. (B, E) Corresponding experimental and simulated extinction spectra in water. (C, F) Simulated and measured extinction of a 2D plasmonic lattice at periodicity (a = 440 nm) for a broad case and (a = 420 nm) for a narrow case. To reflect the particle polydispersity, the simulated spectra were averaged for different particle diameters. Adapted with permission.[4] [ ] Copyright 2019, *ACS Applied Materials & Interface*

2.1.2.   Investigation of the influence of particle size distribution on SLR quality

To investigate the influence of particle size distribution on SLR quality, square lattices of two different particle systems were fabricated: one with a broad particle size distribution and one with a narrow size distribution that shows an eightfold smaller standard deviation (Figure 2A, D). Citrate- and surfactant-stabilized synthesis routes were used to produce broad and narrow size distributions, respectively (see the Experimental Section for more details). The low-molecular weight ligands from syntheses were exchanged against insulin and polyethylene glycol (PEG) to demonstrate the tolerance of the employed assembly technique concerning different surface chemistries. Indeed, the achieved filling rates are comparable, being 81 and 79%, respectively. Moreover, the utilized coatings provide good biocompatibility and help integration in biomedical sensing applications.[109]

Narrowly distributed particles (Figure 8F), the fwhm is reduced from 24 to 16 nm (Q-factor $\lambda/\Delta\lambda$ from 27 to 36) as compared to more polydisperse particles (Figure 8C). This highlights the importance of particle quality, as expected from the corresponding simulated spectra (Figure 8C, F). Note that the mode structure is nicely reproduced by simulations for the narrow size distribution (Figure 8F), whereas the LSPR is only weakly reproduced for the broader case (Figure 8C). The larger size distribution and scattering cross section result in a broad LSPR that is less pronounced as compared to the SLR peak. It is important to mention that in simulations an infinite lattice is considered, whereas an experiment averaging over a 12 mm² area reflects variations in particle coverage. Therefore, the ratio of SLR to LSPR extinction is larger in simulations as compared to experiments.

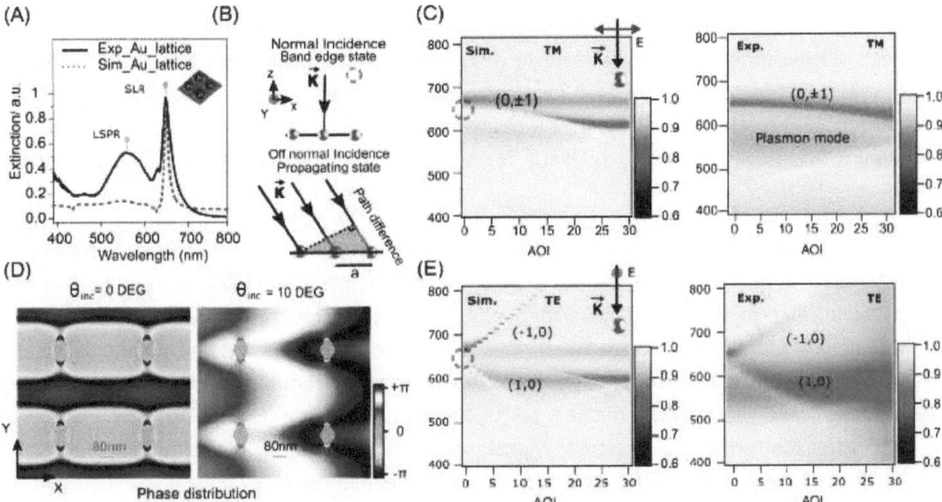

Figure 9. Dispersion relation of a 2D plasmonic lattice: (A) Measured and simulated extinction spectra of gold nanoparticle lattice with normal incidence. (B) Schematic showing the front views of a 2D lattice under normal and off-normal incidence. Simulated and experimental dispersion relation of an index matched plasmonic lattice with periodicity 440 nm under TM (C) and TE (E) polarization. In-plane phase distribution at SLR wavelength for normal and oblique excitation (D). Definition of abbreviations: Angle of incidence (AOI), incident wave vector (K).

We also performed the mean square error (MSE) calculations. The MSE is defined as the average squared difference of the normalized extinction values between the simulation data and what is measured experimentally. The MSE value in the case of broadly distributed gold nanoparticle lattice SLR is 0.11 and for the narrow size distribution it is 0.13. This leads to the fact that experimental spectra match nearly perfectly with simulations. The small oscillation between LSPR and SLR represents the Rayleigh anomaly. This is the wavelength where diffraction turns from being evanescent to radiative.[104] In general, a larger period (440 nm) is expected to produce a narrower SLR as compared to 420 nm because of less overlap with plasmonic losses. The fact that the narrowly distributed particles show a smaller fwhm even for a less optimal 420 nm lattice emphasizes the benefits of monodisperse building blocks. Please note that despite the different synthesis routes used, both particle types are monocrystalline and consequently show comparable plasmonic damping. Furthermore, insulin and PEG coating of particles produce a similar dielectric environment as both represent organic ligands.

2.1.3. Band diagram analysis of 2D plasmonic lattice

The position and line width of the SLR does not only depend on the particle size and periodic spacing. Also the angle of incidence (AOI) can tune the SLR as studied in the dispersion relation diagrams in (Figure 9). In general, a distinction has to be made between normal incidence (so called band-edge SLR modes) and oblique incidence (propagating SLR modes). At the band-edge, all plasmonic dipoles oscillate in phase to form a standing wave (Figure 9D; AOI = 0° case). This feature can be used for single mode in-plane lasing applications.[110] At oblique incidence, the SLR propagates with a phase front, which shows a non-zero group velocity (Figure 9D; AOI = 10° case) relevant for wave front manipulations and phase contrast imaging.[111]

Figure 9C and E show the simulated and experimental SLR dispersion relation, for the cases: (TM) electric field vector in the plane of incidence (x-z-plane) and (TE) electric field vector perpendicular to the plane of incidence. For normal incidence in TM excitation at SLR wavelength 640 nm, all nanoparticles radiate in phase and scatter light perpendicular to the plane of incidence, i.e. along the y-axis. As we move away from the band-edge by increasing AOI, the wave vector (Kx) induces a phase delay in x-direction (Figure 9B), which does not affect the scattering in y-direction. Therefore, SLR follows the (0, ±1) diffraction mode and remains degenerated. In contrast, under TE excitation, nanoparticles radiate in x-direction. As we increase the angle of incidence (vector Kx is non-zero) it provides a phase delay in x-direction. Hence, non-degenerate states of diffraction anomalies (-1, 0) and (1, 0) arise. Hence, this kind of angle dependent response can be tuned over a wide range of wavelengths by compromising the quality factor and coherence.

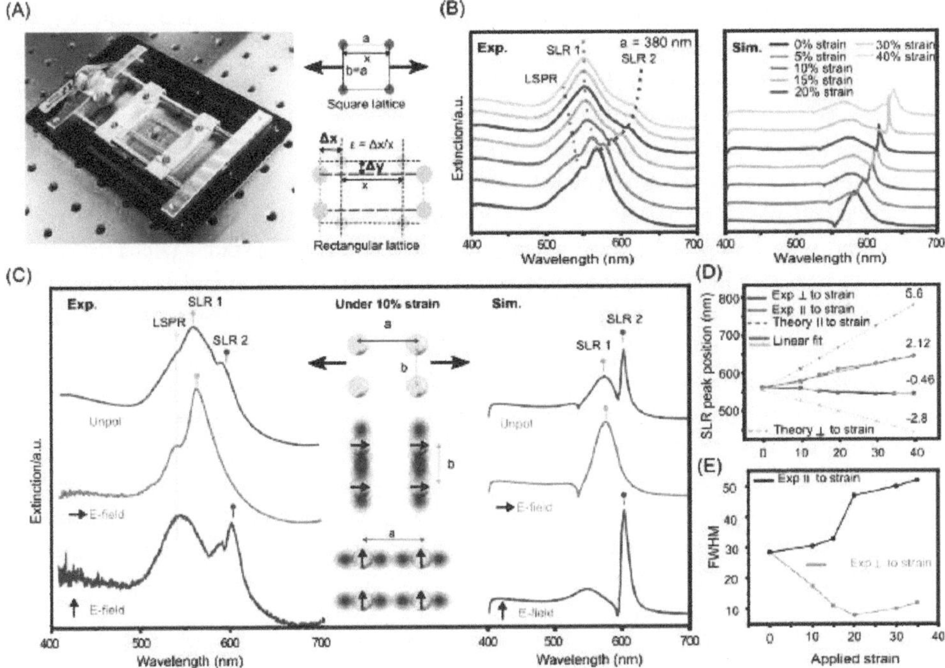

Figure 10. Mechanotunable SLR. (A) Photograph of the stretching device with a clamped sample in the center and schematic of the square lattice under uniaxial strain. (B) Experimental and simulated spectra of the gold nanoparticle lattice under various strains (considering partial force transfer in simulations). (C) Experimental and simulated polarization-dependent selective reading of the lattice resonance modes. (D) Comparison between theoretically expected (100% force transfer) and experimentally obtained SLR peak positions. (E) fwhm of the SLR peaks for various strains. Adapted with permission.[4] [ ] Copyright 2019, *ACS Applied Materials & Interface*

2.1.4. Strain induced tuning of SLR

We discuss a 2D square plasmonic lattice embedded in polydimethylsiloxane (PDMS) that shows a dynamic tuning of lattice constants under mechanical deformation. Because of a Poisson's ratio (which is the negative of the ratio of transverse to axial strain) of 0.5, stretching in the x-direction results in a reduction in the y-direction. Consequently, applying a uniaxial strain transforms the system from a square to a rectangular lattice (Figure 10A). Generally, a non-stretched plasmonic square lattice enables coupling between single particle LSPR and degenerate 2D Bragg diffraction to produce two hybridized modes.[112] Upon deformation into a rectangular lattice, two different Bragg modes (along the x- and y directions) arise, which interact with the LSPR and allow switchable hybrid modes with polarization dependence.

In order to modulate the SLR mode at normal incidence, we benefit from the flexible nature of the UV-PDMS substrate (Figure 10). We discuss here the mechanical tunability only with the narrow particle size distribution particles and compared two different periodicities, 420 and 380 nm. Stretching experiments for broadly distributed particles at a periodicity of 440 nm are provided in Figure 26 of Appendix 6.4. The lattice constant of our particle array can be varied in situ by straining the substrate uniaxially. Because of its centimeter square dimensions, the optical properties of the final plasmonic lattice could be studied using conventional UV–vis spectroscopy with a beam size of 3×4 mm^2. Figure 10A shows the photograph of a home-built stretching device. Increasing the strain gradually from 0 to 40% transforms the square lattice into a rectangular one with lattice constant a (along the applied strain direction) and b (along the perpendicular direction). Two different lattice constants in the x- and y-directions result in two different SLR modes.[86] Because of the strain-induced non-degeneration, one of the SLR shifts toward higher wavelengths (violet color marker in Figure 3B), whereas the other one is blue-shifted (pink color marker) for unpolarized light illumination. We found that the macroscopically applied strain was only partially transferred to the lattice (more details below). Therefore, the simulated spectra displayed in the right panel of Figure 10B were calculated for the effective lattice constraints as derived from the experimental spectra. The experimentally observed spectral shifts are in good agreement with FDTD simulations. In general, SLR1 is broader as compared to SLR2 because of the reduced plasmonic damping of gold for larger wavelengths in the visible wavelength range.[10]

Moreover, the Q-factor of SLR is an interplay between the single particle scattering cross section and the Bragg mode. The low-quality SLR1 shows a significant overlap with the broad LSPR mode for all applied strains. In contrast, the overlap of SLR2 gradually decreases when red-shifting with increasing strain. A strain of 30% (20% in experiments, Figure 10E) represents an optimal balance between LSPR and Bragg modes. The reduced overlap accompanied with less plasmonic character produces a very sharp peak SLR2. When increasing the strain further, SLR2 moves out of the scattering cross section of the single particle, leading to peak broadening. This optimization by adjustment of periodicities of the lattice is explained theoretically in Figure 1. The two different SLR modes in rectangular lattices can be studied in more detail when using linearly polarized excitation, as depicted in Figure 10C for a 10% strain. Note that the SLR induced by polarization in the x-direction is at lower wavelengths as compared to the y-excitation. At first glance, this is counterintuitive as the periodicity in the x-direction is larger. However, this can be understood by considering every plasmonic particle as a dipole antenna. This effect is highlighted in the schematic illustration in Figure 10C. Excitation along the x-axis induces electron oscillations in the parallel orientation for each particle. Yet, these dipoles radiate the scattered light mainly perpendicular to the dipole moment. This concludes that the surface lattice mode always propagates normal to the incident E-field direction. Or, in other words, for an asymmetric lattice, the SLR mode is defined by the lattice constant which is normal to the incident polarization.

2.1.5. SEM and force transfer analysis in 2D plasmonic lattice under various strain

As mentioned before, we see a deviation between macroscopically applied strain and effective deformation of the lattice. Figure 10D compares the strain-induced experimental peak shifts with the theoretical values (considering 100% transfer of applied strain and Poisson's ratio of 0.5). SLR1 is modulated by strain parallel to the stretching direction, whereas the shift in SLR2 reflects the perpendicular strain. The comparison between the slopes (linear fit) of the strain dependent SLR position for theory and experiment gives a measure for partial strain transfer. The ratio between slopes gives 37.8% transfer parallel to the stretching direction and 16.4% in the perpendicular direction (ratio of perpendicular/ parallel 0.43). This is in good agreement with the in situ stretching results in dark-field microscopy which are shown in Figure 30 of Appendix.

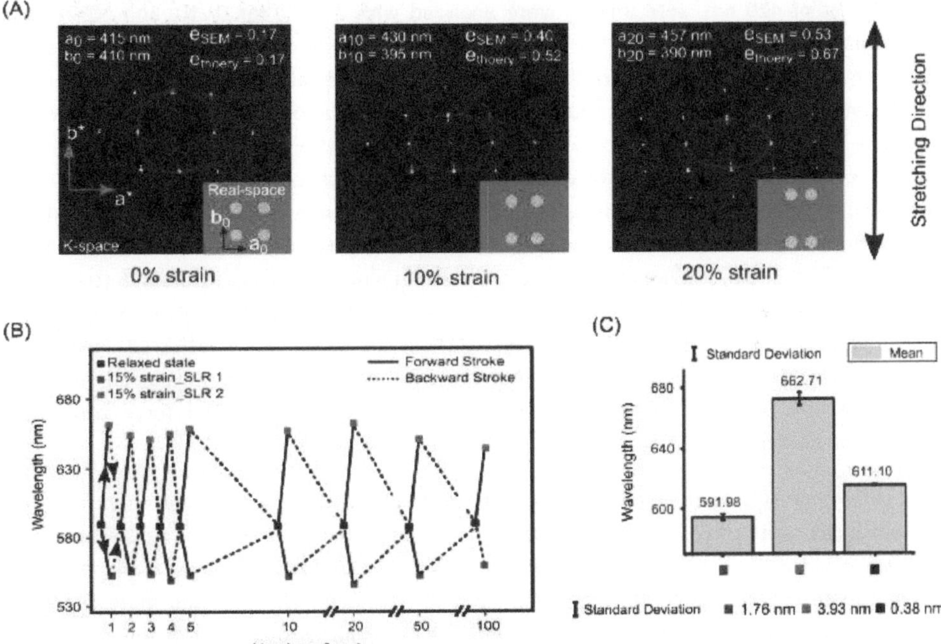

Figure 11. Evaluation of strain transfer. (A) 2D fast Fourier transform (FFT) of 0, 10, and 20% strain SEM images. ai, bi and ai * bi * are the lattice constants in real space and k-space respectively, with i giving the applied strain in percent. (B) Stretch and relax cycles of a plasmonic lattice (a = 420 nm) plotted against SLR peaks for 15% strain. (C) Mean value and standard deviation in SLR peak positions are shown for relaxed and stretched cases. Abbreviation e represents the ellipticity of the ellipse. Adapted with permission.[4] [ ] Copyright 2019, *ACS Applied Materials & Interface*

We attribute the partial transfer of strain to the creep of the polymer chains and deformation of the clamped region when applying a macroscopic strain. It is important to emphasize that the optical quality can be changed by the stretching (Figure 10E). In general, one of the resonance peaks undergoes a red shift, whereas the other one is blue shifted under successively increased strain. As compared to a relaxed state, the coupling strength between LSPR and Bragg modes decreases under strain for the red-shifted SLR modes (narrow fwhm in the case of 30% strain). The blue-shifted SLR successively becomes broader in terms of line width (fwhm) because of increased overlap with the single particle scattering cross section that leads to more plasmonic losses. Depending on the application aspect, this gives freedom in terms of fwhm selection.

The optical analysis shows that the applied macroscopic strain can only be converted by about 38 and 16%. In order to justify this relationship, we have carried out SEM measurements of the strained samples (Figure 11). These measurements were performed for a plasmonic lattice constant of 420 nm. SEM images were analyzed with 2D FFT for 0, 10, and 20% strain. Evaluated lattice constants (a, b) in parallel and perpendicular with respect to the real space strain direction (black arrow) are shown in Figure 11A. In the relaxed state, the FFT shows that the lattice itself is not perfectly square. During the LIL fabrication, the samples are illuminated twice and mechanically rotated 90°. Inaccuracies in the rotation can lead to a deviation in the lattice pattern. We compared the eccentricity values from theory and SEM to understand the deviation of the ellipse from being circular. Eccentricity values show a clear trend that UV-PDMS grating was stretched but the force transfer on to the particle grating was partial. In the case of 20% strain, the lattice constant in the strain direction changes from 415 to 457 nm. This translates to a 50% strain transfer as compared to theoretical values. Similarly, for perpendicular to the strain direction, only 24% of the strain was transferred. Finally, the stretching inside the SEM confirms that there is moderate force transferred from the macroscopic scale to the nanometer lattice deformation scale. The force cannot be implemented one to one probably because of the clamp holder.

To stretch the PDMS, the material is squeezed at the clamp holder with an unrecognized force, which can cause the mismatch. Note that the deviation between SEM and optical spectrum data could also be because of the absence of the index matching layer during the SEM measurement. Non-uniform embedding leads to less frictional forces and more degree of freedom in terms of particle motion. In order to confirm the robustness of the elastomeric plasmonic lattice, the sample was stretched and relaxed 100 times and the bright-field spectrum was measured after each cycle (Figure 11B). The maximum strain was kept to a constant value of 15% for all the cycles. The black color dots (relaxed state) in Figure 11B show that UV-PDMS recovers even after 100 cycles with a standard deviation of only 0.38 nm, as summarized in Figure 4C. The deviation in the values for SLR1 (blue dots) and SLR2 (red dots) is higher as compared to the relaxed state. This can be correlated with the manual handling of the stretching device.

Conclusion

In conclusion, we discussed the assembly of ordered nanoparticle arrays that show a narrow and mechanically tunable SLR over 100 cycles of deformations. The assembly relies on directed self-assembly of colloidal nanoparticles in templates formed by soft lithography. We showed the arrangement with broad/narrow particle size distribution and discussed their influence on the optical quality. The real-time mechanical tuning and accessibility of non-degenerated SLR modes hold great potential for plasmonic lasing,[113] biosensing,[114] or colorimetric sensors.[84] We show that the optical properties of the system change dramatically upon applying strain, as the array changes from a square lattice to a rectangular lattice. Importantly, the optical quality is not lost upon straining the system and the SLR peaks can be shifted over a wavelength range of 70 nm. Moreover, the uniaxially strained lattice opens up the possibility for introducing optical anisotropy. This scalable system with monocrystalline nanoparticle building blocks provides a platform to curb the rigid nature of optical systems and offers versatile plasmon mode engineering within flexible electronics. Finally, the rational design framework established also allows for extending the approach to supracolloidal structure (clusters, hetero-clusters, etc.).

Methods

Template Fabrication. Nanopillars with various periodicities and diameters were fabricated with LIL using Lloyd's mirror setup. Further experimental details can be found in Figure 22 of Appendix 6.1. A cleaned silicon wafer substrate (2.54 cm × 2.54 cm) was first spin-coated with a back-antireflection coating AZ BARLi II (70 nm in thickness). Further, a layer (214 nm thickness) of AZ MIR 701 14cp (Micro Chemicals GmbH) positive resist, 1 to 1 ratio diluted with ethyl L-lactate, was spin-coated (6000 rpm, 60 s) on top. The sample was exposed twice to the laser with 90° sample rotation. A wet chemical development process was used to etch the exposed resist in order to fabricate the pillars (master structures). The sample was immersed in AZ 726 MIF developer (Micro Chemicals GmbH) for 30 s followed by gentle rinsing with water and nitrogen drying. Soft lithography was employed to fabricate elastic molds of the as-prepared periodic nanostructures using silicone rubber PDMS, called UVPDMS. We used two-component KER-4690 photo-curable PDMS (Shin Etsu & Micro Resist Technology GmbH) with a 1 to 1 mixing ratio. The mixture was stirred for 30 min through the magnetic stirring process and degassed for 20 min. The blend was drop-casted on the substrate, degassed again, and cured with a 365 nm UV lamp for 30 min. After the peel-off process possible remains of the photoresist were removed by rinsing with ethanol and drying with nitrogen.

Nanoparticle Synthesis. Particle Synthesis of Broad Size Distribution. Citrate-stabilized nanoparticles were prepared via seed-mediated growth synthesis.[115] The protein coating was prepared according to earlier published methods.27–29 Briefly, 40 mL of the nanoparticle solution was added to a 4 mL 1 mg mL−1 insulin solution with 1% wt sodium citrate at pH 9 to generate the desired protein coating. The solution was allowed to incubate overnight. After fourfold centrifugation, the nanoparticles were analyzed with UV–vis and Libra 120 transmission electron microscopy (TEM). The nanoparticle size of 83.1 ± 12.2 nm was determined by statistical analysis of 150 particles. Particle Synthesis of Narrow Size Distribution. Spherical gold nanoparticles capped by hexadecyltrimethylammonium chloride were synthesized by a seed-mediated growth process.11 The resulting particles had a size of 76.9 ± 1.5 nm as determined by (TEM) statistics (at least 100 particles). Briefly, the PEG-stabilized particles were synthesized in three steps. First, so-called Wulff seeds are produced by reducing tetra chloroauric acid (HAuCl4) with sodium borohydride (NaBH4) in the presence of hexadecyltrimethylammonium bromide (CtaB). The obtained 2 nm large and single crystalline particles are successively grown by two further synthesis steps until they reach the desired particle size. During the growing process, HAuCl4, ascorbic acid, and CtaC serve as the Au precursor, reductant, and stabilizing agent. In order to guarantee kinetic control and thus control over the particle shape, a syringe pump system is used for the last growing step. The resulting particles were purified by centrifuge washing and set to a surfactant concentration of 2 mM. For the self-assembly step, a ligand exchange was performed from CtaC to PEG-6k-SH.30.

Template-Assisted Colloidal Self-Assembly. Prior to the assembly experiment, the microscopy glass slide (Menzel)was hydrophobized via gas-phase deposition of trichloro (1H, 1H, 2H, 2 Hperfluorooctyl) silane (448931, Sigma-Aldrich) at 60 °C for 3 h. Using a custom-made setup, 15 µL nanoparticle suspension [0.5 mg mL−1 gold, 0.25 mM sodium dodecyl sulfate, 0.025 wt % Triton X-45 (Sigma-Aldrich)] was confined between the template and the stationary glass slide (gap ≈ 500 µm) and the patterned substrate was withdrawn underneath the droplet at a speed of 1 µm s−1 by a motorized translational stage (PLS-85, Physik Instrumente). Whereas the surfactants added tune surface tension and contact angle to enable selective deposition of the particles,19 the elevated solution pH value (pH 9) in the case of insulin coating ensures colloidal stability by electrostatic repulsion. The experiment was carried at 12 K above dew point. Further experimental details can be found In Figure 28 of Appendix 6.6. Prior to optical characterization, the samples were index-matched by spin-coating UV-curable PDMS at 2000 rpm for 60 s.

UV–Vis–NIR Spectroscopy. Optical spectra were measured on a Cary 5000 spectrometer (Agilent, USA) using the Cary universal measurement accessory. The spot size was fixed to 3 × 4 mm2 for UV–vis and NIR detectors. All the dispersion spectra were measured by rotating the sample plane by keeping the plane of incidence constant.

Bright-Field Imaging. Images of the self-assembled nanostructures were captured by a Nikon Eclipse LV100ND microscope using a bright-field reflection microscopy setup. The surface was illuminated by a halogen light source (12 V, 50 W) focused by a 5× objective lens. Images were taken by a Nikon DS-Fi2 5.24-megapixel charge coupled device (CCD) camera and analyzed by Nikon imaging (NISElements) software. Unpolarized light was used for all the images.

Bright-Field Spectroscopy. Spectroscopy under strain was performed with a home-built stretching device on a transmission microscope (Nikon Ti-U). The microscope was equipped with an IsoPlane-160 spectrometer and a PIXIS: 256 CCD camera (Princeton Instruments). The light beam (100 W halogen lamp) was focused with a bright-field condenser (Ti-C-LWD) and collected with a 60× air objective (CFI S Plan Fluor ELWD, NA 0.7). Spectra with 383 nm bandwidths were taken at 600 and 750 nm center wavelengths for 0.5 s (New Technologies and Consulting/Harald Giessen, NT&C software). All spectra were corrected by subtraction of the CCD's dark current and normalized by a reference spectrum, which was recorded in an area without particles.

Spectroscopic Ellipsometry. RC2-D dual rotating compensator ellipsometer, ex situ with automated tip and tilt alignment stage from J.A. Woollam Ellipsometer Solutions was used for refractive index measurements. Further, measured data were fitted with Complete Ease software via the Cauchy substrate model.

FDTD Simulations. We have used a commercial software package for performing FDTD simulations (Lumerical FDTD, version 8.16).31 For the simulation of the optical response, a plane wave source was used and the frequency points were set to be half the wavelength span. Two-dimensional (2D) frequency-domain fields and power monitors (transmission and reflection monitors in the direction parallel to the lattice plane) were used to obtain the optical responses of the system. For the dielectric properties of gold, data from Johnson and Christy were fitted using six coefficients with an rms error of 0.2.32 The refractive index of the substrate (UV-PDMS) was measured via spectroscopic ellipsometry (Figure S6). In order to determine the electric field and surface charge density, we simulated the model at the LSPR and hybrid mode (SLR) frequencies. Periodic boundary conditions were used for square plasmonic lattice simulations in the lattice plane direction. For stretching, periodicities of the lattice in both the directions were increased based on the tabulated Poisson's ratio. All simulations reached an auto-shutoff of at least 10−7. For the best simulation stability, the mesh area was chosen to be at least 100 nm larger than the existing structure in the lattice plane direction.

Atomic Force Microscopy. For evaluating the topography of photoresist masters and PDMS molds, atomic force microscopy (AFM) height images were measured using a Dimension 3100 NanoScope V (Bruker, USA) operated in tapping mode. Stiff cantilevers (40 N m−1, 300 kHz, Tap300, Budget Sensors, and Bulgaria) were employed.

Scanning Electron Microscopy. A NEON 40 FIB-SEM workstation (Carl Zeiss Microscopy GmbH, Oberkochen, Germany) operated at accelerating voltage (electron high tension) of 1 kV was used for capturing scanning electron micrographs.

2.2. Hybridized Guided-Mode Resonances via Colloidal Plasmonic Self-Assembled Grating

This section is based on the peer reviewed journal article. Adapted under the terms of a ACS Author Choice with Creative Commons Noncommercial No Derivative Works (CC-BY-NC-ND) Attribution license.[36] [ ] Copyright 2019, *ACS Applied Materials & Interface.*

By *Swagato Sarkar,* **Vaibhav Gupta,** *Mohit Kumar, Jonas Schubert, Patrick T. Probst, Joby Joseph and Tobias A.F. König**

Swagato Sarkar, Vaibhav Gupta, Mohit Kumar, Jonas Schubert, Patrick T. Probst, Tobias A.F. König
Institute for Physical Chemistry and Polymer Physics, Leibniz-Institut für Polymerforschung Dresden e.V. (IPF), Hohe Str. 6, 01069 Dresden, Germany.
Mohit Kumar, Swagato Sarkar, Prof. Joby Joseph
Photonics Research Lab, Department of Physics, Indian Institute of Technology Delhi, 110016 New Delhi, India.
Tobias A.F. König
Cluster of Excellence Center for Advancing Electronics Dresden (cfaed), Technische Universität Dresden, 01062 Dresden, Germany.

* Corresponding author

Author contribution statement

VG has performed all the template assisted self-assembly experiments with the guidance of PP. JS synthesized the required nanoparticles. SS, **VG**, MK did the template fabrication. SS has performed the software-based (FDTD) simulations. SS and **VG** carried out the sensing experiments. **VG** encouraged MK and SS to investigate the simulation and optical spectroscopy methods as well as supervised the execution and findings of this work. JJ and TK contributed in developing the concept and writing of **the manuscript.** All authors provided critical feedback and helped shape the research, analysis **and manuscript.**

Introduction

Optical antennas are the key to confine light energy into a target structure and vice versa.[5] Plasmonic nanoparticles (NPs), in particular, are suitable for collecting light energy into a nanometer-sized volume via oscillation of the free electrons at the particle surface (localized surface plasmon resonance, LSPR) and converting the energy into a current, that is, light harvesting.[116] Thus, efficient light concentration and hot carrier extraction have led to diverse applications such as photocatalysis,[117] photovoltaics,[118] and photodetection.[119] Plasmonic NP chains are of special interest as they can excite modes with suppressed radiative losses.[120,121] When a chain is formed from individual NPs, the excited plasmon mode for incident polarization along the chain direction shifts to a lower energy (super-radiant mode) until a certain group velocity of the collective plasmon mode is reached.[122] This collective plasmon mode along with all higher energy modes can be imagined analogously to a mechanical string vibration model or non-degenerated modes as in quantum mechanics. In theory of plasmonic chains, these energetic higher modes (subradiant) have lower propagation losses than the super-radiant mode.[123] Recent research topics have focused on the study of these subradiant modes for possible applications in optical sensing,[124] sub-diffraction energy transport,[125,126] and plasmonic NP gratings.[127] However, these NP chain-based plasmonic grating resonant (PGR) modes still have radiative damping, and their resonance wavelength varies greatly because of size and particle spacing.[127] By incorporating thin-film layers, plasmon modes can be extracted by phase-matching conditions and guided within these layers following the properties of a grating waveguide-based guided-mode resonant[128,129] (GMR) structure. Such waveguide–plasmon polaritons have been intensively studied and can be used to investigate strong coupling,[130–132] to demonstrate light generation,[133] and to show filter properties with high optical transmission.[134] All these properties benefit from the hybridized coupling (Fano resonance) of a narrow (waveguide mode) and a broad (plasmonic mode) oscillation.[135] In most of the cases, these hybrid structures are fabricated using an electron beam and contact lithography followed by physical vapor deposition, sol–gel approach, and a chemical etching step which results in a limited array of polycrystalline NPs that suffer from increased damping. Recent developments in directed self-assembly can solve this bottleneck to realize large-scale assembly of monocrystalline NPs at a reasonable price.[45] In particular, the combination of rapid top-down laser interference lithography (LIL) template and bottom-up template-assisted self-assembly (TASA) methods allow the freedom to assemble NPs in various two-dimensional structures.[136] Consequently, development in self-assembly makes it possible to access the plasmon modes along the NP line selectively. Generally, the NP-based pure plasmonic resonance mechanisms are extremely responsive to the change in the surrounding media and thus have significantly contributed toward the fabrication of highly sensitive refractive index[137] (RI) as well as biological sensors.[138,139] As broader PGR (super- and subradiant) modes with a larger full width at half maximum (fwhm) have a higher radiative loss, Q-factor (inversely proportional to fwhm) remains significantly low for their application in the field of sensing. By coupling the plasmonic features to the high "Q" photonic modes (with smaller fwhm) of a dielectric GMR structure, one can increase the sensitivity of the so-obtained hybridized photonic modes as compared to the pure photonic (GMR) case while maintaining the sharpness and "quality". In this work, we have applied plasmonic grating, that is, the NP lines (fabricated by LIL and TASA) directly to the waveguide (TiO2 layer) to save the process steps in constructing the hybridized geometry. A previous work from our group has demonstrated plasmonic studies of NP lines formed by template-assisted colloidal self-assembly, followed by a printing transfer to the target substrate.[127] Here, we save ourselves from the transfer step by assembling the particles directly into the GMR structure. However, this step is not trivial as the surface chemistry of the particles and the substrate required for an efficient assembly now had to be adjusted to the LIL-fabricated GMR structure. Moreover, this effective fabrication step was not possible in the previous work as those substrates were made by mechanical instability (wrinkling) followed by assembly through spin-coating. For the current work, we have used LIL and convective self-assembly toward the fabrication of the hybrid opto-plasmonic structures which till now are limited by "e-beam" fabrication techniques.[130,131] The grating period and waveguide thickness are chosen accordingly to obtain a phase match to favor coupling between the GMR and PGR modes. Experimental and theoretical spectroscopic methods with polarizations both parallel and perpendicular to the chain direction are used to study the plasmonic NP–waveguide polaritons on the centimeter scale. With this simple, cost-efficient, and upscalable fabrication method, we are able to discuss the complete hybrid nature of the resulting modes and show their application as opto-plasmonic RI sensors.[140]

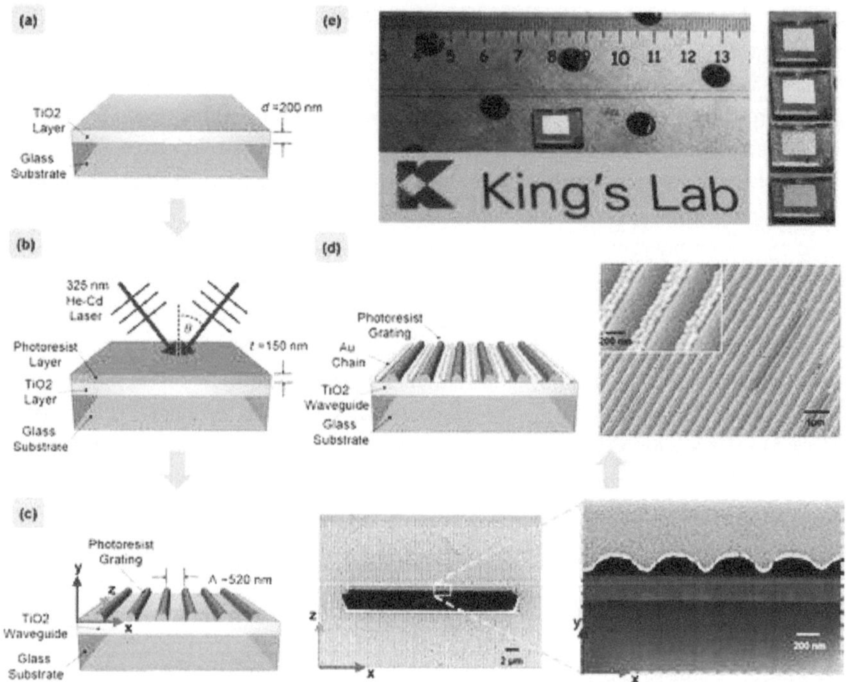

Figure 12. Fabrication of the hybrid opto-plasmonic structure: (a) deposition of 200 nm TiO2 layer onto the glass substrate by electron beam physical vapor deposition. (b) Spin-coating of negative photoresist and LIL using a He–Cd laser. (c) Formation of a GMR structure with a 520 nm periodic photoresist grating. Focused ion beam (FIB) cut reveals grating channels in a magnified view. (d) Directed self-assembly of gold NPs (88 nm in diameter) by the controlled evaporation of particle dispersion (convective self-assembly). An SEM image of the final structure with a magnified view in the inset shows dimer particle chain arrangement within grating channels. (e) Photograph of the fabricated sample on an optical table. Different first-order diffracted wavelengths (colors) at various AOIs appear on reflections from ordinary white room-light sources. Adapted with permission.[36] [ ] Copyright 2019, *ACS Applied Materials & Interface*

2.2.1. Fabrication of hybrid opto-plasmonic structure via template assisted self-assembly

Figure 12 describes the overall fabrication procedure of the desired hybrid geometry, which is a combination of top-down and bottom-up approaches in producing the dielectric GMR structure and plasmonic NP grating, respectively. Once the photoresist grating of the GMR structure is formed, the grating channels are filled with synthesized gold NPs through the process of directed self-assembly (TASA). Thus, one can observe a fast filling process of gold

NP arrangement in chains over a large area which may be considered superior in comparison to the electron beam deposition techniques. Although grating homogeneity was lacking over a large scale that resulted in partial filling of some grating channels, the overall characteristics remain maintained when compared to the simulative study of ideal homogeneous chains and are discussed in the following sections. Figure 12d shows the finalized structure with an SEM image exhibiting efficient filling of photoresist channels with gold NPs. The magnified view in the inset supports the formation of dimer chains, which is considered during all simulative analysis throughout this report. Thus, our method, in contrast to e-beam techniques provides not only cost efficiency and large area fabrication, but also a plausible effort toward bringing in both the physical and chemical communities to a single productive platform. Moreover, plasmonic NP chains have been directly assembled in different designated geometries for studying both pure and hybrid states. The non-necessity of any molded template and transfer mechanism, unlike previous study,[127] has indeed made the whole fabrication process even simpler.

Further in the same direction Sarkar et al. studied the same behavior for metal (Au) nanoparticle chains and claimed that waveguide plasmon polariton can be excited in transverse electric (TE) polarization in contrast to the metal bar case. We further go into more details of the mechanism of this phenomenon.

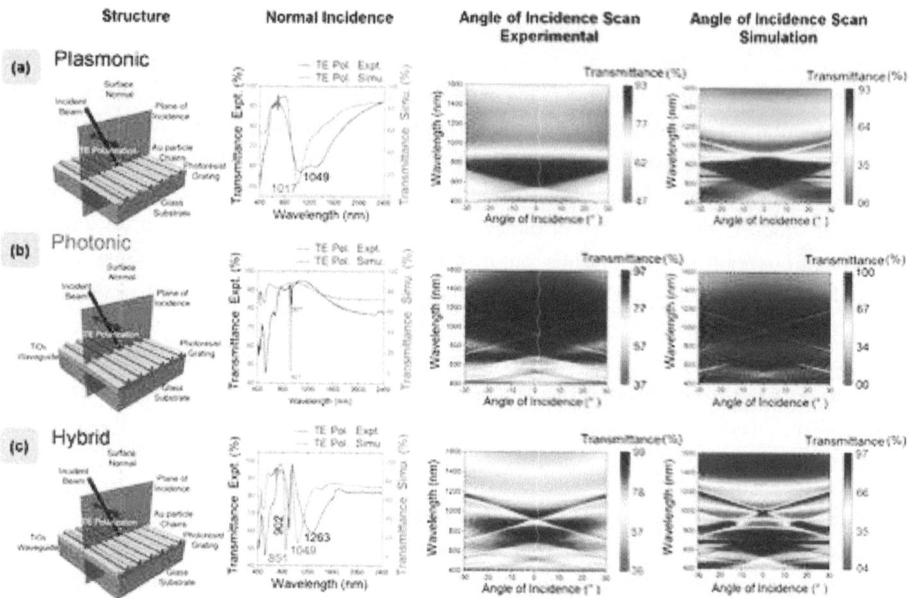

Figure 13. Comparison of optical properties of three different structures in TE mode: (a) plasmonic dimer lines of gold NPs with 520 nm periodicity supported by photoresist gratings on a glass substrate. Both experiment and simulation studies show a broad particle resonance around 1000 nm for excitation with TE polarization along the chain directions. Variation of the incidence angle experimentally exhibits a constant broad dip which is also supported by simulations. (b) Dielectric GMR structure with a photoresist grating (of periodicity 520 nm and thickness 150 nm) on a TiO2 waveguide (of 200 nm thickness) and a glass substrate. Experimental as well as simulation studies observe transmission dip at 921 nm for normal incidence, whereas splitting of modes occurs on varying the AOI. (c) Hybrid plasmonic–dielectric resonant structure with gold NP dimer lines filled within the grating lines of a dielectric GMR. At normal incidence, two hybrid modes are excited at different wavelengths. Scanning of the incidence angle reveals the interaction of the broad plasmonic mode with the photonic GMR mode at higher angles, as observed from both experimental and simulation data. Adapted with permission.[36] [ ] Copyright (2019), *ACS Applied Materials & Interface*

2.2.2. Comparison of optical band diagram of three (plasmonic, photonic and hybrid) different structures in TE and TM modes

Figure 13a presents studies on such plasmonic blocks excited in parallel polarization, that is, the TE case with respect to the plane of incidence lying normal to the grating lines (shown in the figure). Under normal incidence in air, the experimental spectroscopic data exhibit collective plasmonic resonances around 1049 nm as a broad transmission dip that well-agrees with previously reported findings[127] and can be identified as the super-radiant mode. For a better understanding of the associated resonance, a simulation model with similar geometry of gold NP dimer chains on a photoresist patterned glass substrate is considered using FDTD methods, with the specifications given in the Experimental Section. The deviation between the experimental and theoretical transmittance values can be attributed to the non-ideal cases of finite chain length along with the missing particles in reality as compared to the modeled ideal case of infinitely chained close packed structures. Further, studies under the variation of AOI are performed. As the polarization (TE) is fixed parallel to the chain orientation, varying the incident angle keeps the incident electric field intact, thus producing similar transmission spectra over −30° to 30°. The transmittance values as a function of wavelength and incident angle obtained experimentally as well as through simulation are provided in Figure 13a that correspond to the dispersion relation. The effects of polarization perpendicular to the grating lines (TM) are also explored via experiment as well as simulation and presented in Figure 13a. It can be observed that plasmonic bands are formed around 530 nm corresponding to a single NP resonant dip. Next, in Figure 13b, dielectric GMR with a photoresist grating and TiO2 waveguide on a glass substrate is studied for optical characterization as well as for comparison with the final hybrid structure. The experimental and simulation transmission spectra match well with a periodicity of 520 nm, waveguide thickness of 200 nm, and a grating amplitude of 150 nm. For normal incidence in air, both the graphs produce a transmission dip at 921 nm for TE polarization, that is, parallel to the grating. However, in contrast to the simulation, the resonant dip in the experimental curve has a shorter depth because of the lesser diffraction efficiency of the fabricated grating. This can be caused by the non-uniformities like the variation in duty cycle (the ratio between the ridge width and the period of grating) and depth over a large area unlike the ideal uniform case of simulation. As the incident angle is varied from 0° toward ±30°, we observe a similar splitting of the resonant mode into non degenerate modes.[128] The purpose of these AOI scans is to observe the effect of the red-shifted order interplaying with the plasmonic band in the final hybrid opto-plasmonic structure. Figure 13b shows similar GMR optical characteristics in the TM mode with a resonant wavelength around 870 nm under normal incidence. Finally, Figure 13c shows the case of a hybrid structure achieved through the combination of the fabrication processes discussed in this report with 150 nm grating thickness, 520 nm grating periodicity, and 200 nm TiO2 waveguide layer supported on a glass substrate. The normal incidence spectrum obtained experimentally (black curve) in air contains two resonant dips: the broader one around 1263 nm as a hybrid plasmonic resonant mode along with a sharper hybrid photonic mode at 902 nm corresponding to the GMR phenomenon. The simulation results (red curve) similarly show the presence of both hybrid plasmonic and photonic resonances at 1049 and 851 nm, respectively. The experimentally obtained GMR transmittance dip (photonic) is enhanced in the hybridized photonic state at 902 nm because of the effect of coupling. As AOI is increased from 0° toward ±30°, one of the photonic modes (red-shifted) starts interplaying more with the constant plasmonic band as their spectral positions get closer. Thus, at larger angles, more of the plasmonic features are induced to the photonic mode through an easy exchange of energy, making both the hybrid modes broad as seen from the figure. The simulation results of the AOI scan in Figure 13c also agree to this experimentally observed fact. This hybridization, however, is strictly a feature of the TE modes as no such case is obtained for the TM mode.

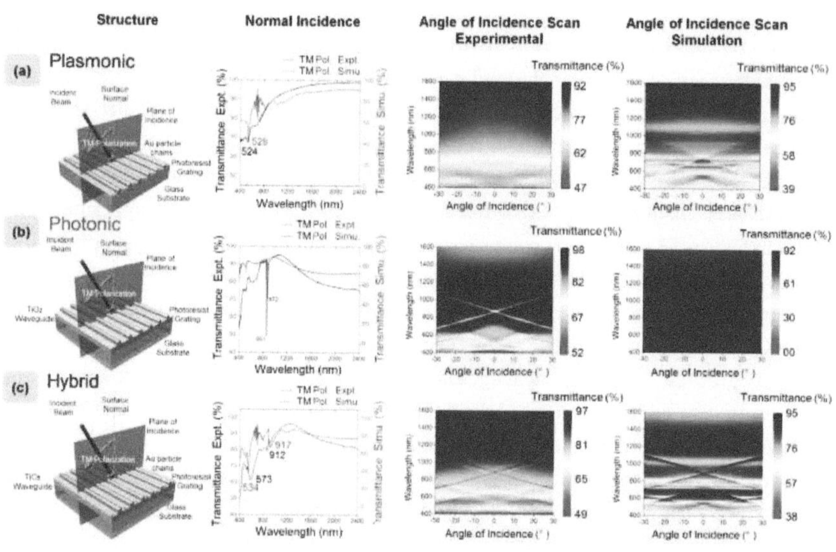

Figure 14. Comparison of optical properties of three different structures in TM mode: (a) Plasmonic dimer lines of gold nanoparticles with 520 nm periodicity supported by photoresist gratings on a glass substrate. Both experiment and simulation studies show particle resonance around 530 nm for excitation with TM polarization across the chain directions. Variation of the incidence angle experimentally exhibits a constant dip supported by simulations. (b) A dielectric GMR structure with photoresist grating of (periodicity 520 nm and thickness 150 nm) on TiO2 waveguide (of 200 nm thickness) and the glass substrate. Experimental studies observe transmission dip at 872 nm for normal incidence and similar dip at 861 nm for simulation with the splitting of modes on varying the angle of incidence. (c) A hybrid plasmonic-dielectric resonant structure with gold nanoparticle dimer lines filled within grating lines of a dielectric GMR. On normal incidence, particle modes are excited around 550 nm with guided waveguide modes at far separated higher wavelengths of around 915. Scanning of incidence angle reveals non-interaction of plasmonic mode with the photonic GMR modes angles as observed from both experimental and simulation data. Adapted with permission.[36] [ ] Copyright (2019), *ACS Applied Materials & Interface*

This is because of the fact that for the TE case, the GMR sharp resonance falls on the left shoulder of the broad PGR mode at normal incidence providing a fair chance of the coupling mechanism. In contrast to the optical characteristics of different resonant geometries under transverse electric (TE) polarization as studied in Figure 13, the effect of the transverse magnetic (TM) polarization is also studied here for comparison. Figure 14 presents both experimental and simulative results of the plasmonic, photonic and hybrid structures for the similar scan range (0° to ± 30°) as shown in Figure 13. Plasmonic bands (red) as obtained from the angle of incidence scan shows isolated particle features for the plasmonic chains around structure Figure 14b, the resonant modes (θ_{inc} = 0° and ≠ 0°) doesn't lie within the spectral position of the plasmonic band. Thus the hybrid geometry Figure 14c exhibits no such hybridization for the case of TM polarization.

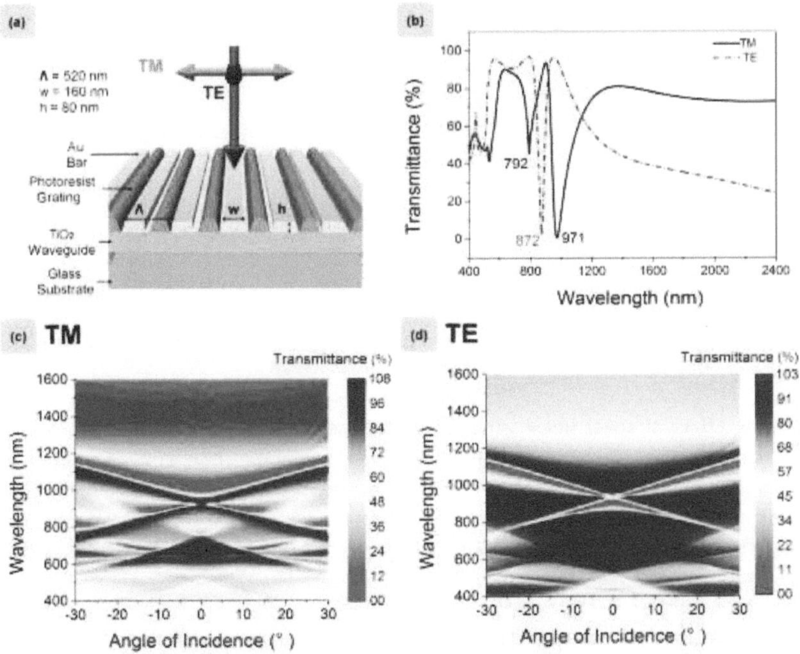

Figure 15. Simulative study on optical properties of hybrid opto-plasmonic geometry with a grating of metallic gold bars: (a) Plasmonic gold nanograting with 520 nm periodicity, 160 nm width and 80 nm height along with photoresist grating on 200 nm thick TiO2 waveguide layer. (b) Transmission spectra of the structure in TM and TE modes under normal incidence. Hybridized modes are obtained in TM polarization at 792 and 971 nm whereas sharp dip relating to GMR is obtained for the TE case. (c), (d) Angle of incidence (AOI) scan with TM and TE polarization respectively. TM mode shows signatures of hybridization similar to that in longitudinal excitation (TE) of plasmonic nanoparticle chains as discussed in the main article. TE mode shows prominent GMR properties of mode splitting with broadening of dips on increasing the AOI. Adapted with permission.[36] [ ] Copyright (2019), *ACS Applied Materials & Interface*

In conclusion TM case as shown in Figure 14, the plasmonic resonances near 530 nm are the effective contributions of single NPs. As located far away from the photonic resonances (around 870 nm), non-overlapping of these modes results in non-coupling, making the TE excitation case the only possible way of hybridization. This is contradictory to the case of well-known waveguide–plasmonic structures[131,132] consisting of metallic bar-based grating where plasmonic modes are only excited in the TM polarization, whereas the TE case produces only a grating-like response.[141]

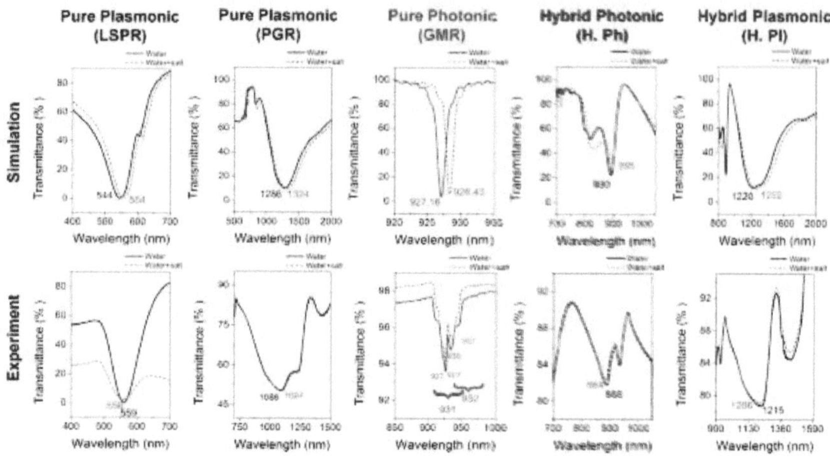

Figure 16. Effect of cover index variation with water as a cover medium: Comparison of both simulation and experimental transmission spectra for geometries operating on different resonance regimes. Deuterium Oxide (D2O) is chosen as a background medium whose refractive index (RI) is varied by mixing sodium chloride (NaCl). For simulation, such media are modeled using RI data (see figure S8 in **publication VG2)** Adapted with permission.[36] [ ] Copyright (2019), *ACS Applied Materials & Interface.*

2.2.3. Simulative comparison of optical properties of hybrid opto-plasmonic NP chains with a grating of metallic gold bars

To compare our plasmonic grating of NP chains with metallic bars. a similar dimension of the grating in the hybrid opto-plasmonic geometry for both the polarizations is studied and shown in Figure 15. The plasmonic gratings of gold bars have been reported to show plasmonic features exciting with polarization perpendicular to the grating lines (TM) whereas normal grating like behavior with parallel polarization (TE). When combined with a waveguide, the TM mode causes generation of waveguide-plasmon polaritons due to strong coupling whereas TE mode results only in the generation of photonic waveguide modes leading to GMR. We compare our hybrid geometry by replacing nanoparticle chains with nano bars where the height and width are kept identical in both cases Figure 15a. One can observe hybridization for the case of TM polarization Figure 15b in case of metallic bars which is expected from literature. This is because, for TM polarization, the plasmonic band of the metallic bar is around 800 nm which can interact with the photonic GMR TM mode resulting in hybridization. For TE case, the photonic mode finds no plasmonic counterpart to interact and thus is devoid of any such coupling. The angle of incidence scan further confirms the outcome as seen from Figure 15(c-d). This points out that unlike our case, TM excitation instead of TE is indeed required to bring in the hybridization for these types of waveguide–plasmonic structures.

2.2.4. Effect of cover index variation with water as a cover medium

A detailed study on the RI sensing behavior of these pure and hybrid resonant states has been carried out to imply the importance of the hybridized modes using pure water and salt solution. Appendix 6.10 contains RI data (see Figure 34) corresponding to the liquids used. For simulation, these RI data are incorporated into the simulation model as surrounding media to obtain spectral feedback for the different resonant geometries. For the experimental realization, three separate custom-made poly(dimethylsiloxane) cuvettes with distinct samples operating in PGR, GMR, and hybrid modes are introduced to the spectrophotometer, and sensing measurements are performed in response to pure D_2O and salt solution. For a comparative isolated particle LSPR measurement, gold NP solutions in glass cuvettes with different suspension media (pure D_2O and salt solution) are used. Figure 16 contains the sensing data for both the simulation and experiment of all the resonant geometries with water as a cover medium. Sensitivity (S) and FOM are calculated for the corresponding resonant line shapes and are tabulated in Table 2. There is a certain mismatch in the "S" values between simulation and experiment for the cases of plasmonic (both pure and hybrid) modes, which is due to the nonconsideration of the protein coatings around the NPs in the simulation. These coatings, in reality, have affected the experimental plasmonic sensitivity and can be eliminated in future by physical means like plasma cleaning or heat treatment. The photonic modes (both pure and hybrid) remain unaffected, showing matching values for experiment and simulation.

Thus, we can conclude that the sensitivity of the hybridized states is solely dependent on that of the pure states and the medium these pure states are sensitive to. With a proper optimization and choice of these pure states, one can tune between the sensitivity and FOM over a wide spectral range. Thus, a pre-optimized GMR device with an improved "S" (optimization of duty cycle and index contrast) can surely have a further enhanced sensitivity as well as a higher FOM (FOM = S/fwhm) in the hybrid photonic mode as long as the resonant wavelength overlaps with the plasmonic resonance.

Table 2. Refractive index sensing with water: Calculation of sensitivity (S) and figure of merit (FOM) for different resonance geometries from the simulation as well as experimental transmission spectra plotted as a function of different refractive index in Figure 16. The values of the refractive indices at different resonant wavelengths for both the cases of pure water and salt solution are estimated from the Cauchy relation and coefficients given in figure S8 of **publication VG2** Adapted with permission.[36] [ ] Copyright (2019), *ACS Applied Materials & Interface.*

	Method	λ_1 (nm)	n_1 (RIU)	FWHM 1 (nm)	λ_2 (nm)	n_2 (RIU)	FWHM 2 (nm)	Δn (RIU)	$\Delta\lambda$ (nm)	FWHM Avg (nm)	S (nm/ RIU)	FOM (S/ FWHM)
LSPR	Simu.	544	1.3294	88	554	1.3748	90	0.0454	10	89	220.26	2.47
	Expt.	559	1.3289	74	556	1.3747	66	0.0458	3	70	65.50	0.94
PGR	Simu.	1286	1.3213	449	1324	1.3645	474	0.0432	38	461.5	879.62	1.91
	Expt.	1088	1.3221	393	1097	1.3656	393	0.0435	11	393	252.87	0.64
GMR	Simu.	927.18	1.3230	1.52	928.43	1.3667	1.36	0.0437	1.25	1.44	28.60	19.86
	Expt.	931.0	1.3229	21.36	932.0	1.3668	23.19	0.0439	01	22.28	22.78	1.02
H. Ph.	Simu.	890	1.3233	114	895	1.3671	105	0.0438	05	109.5	114.16	1.04
	Expt.	888	1.3233	109	884	1.3673	111	0.044	04	110	90.9	0.82
H. Pl.	Simu.	1220	1.3215	406.5	1252	1.3648	440.5	0.0433	32	423.5	739.03	1.74
	Expt.	1215	1.3219	379	1206	1.3650	368	0.0431	09	373.5	208.81	0.56

Conclusion

In summary, we have demonstrated a new method of combining top-down (LIL) and bottom-up (colloidal synthesis and TASA) approaches for the successful fabrication of hybrid plasmonic–photonic geometry over a large centimeter-scale area. The fabricated structure with a proper choice of design parameters supports coupling of the plasmonic radiant modes of gold NP grating to the photonic modes of TiO2-based GMR under normal incidence; this results in the formation of hybridized states that can be tuned by varying AOI. The hybridized structure is optically characterized in comparison to its constituent resonant geometries along with an in-depth study of these resonances through numerical simulations to understand the coupling of two resonances. On the basis of the electric field plots, a matrix transfer model is also established that can be applied in general to similar cases of coupling of different resonances. With a system completely new in the field of modal strong coupling, we have realized its potential application in RI sensing where the sensitivity of a GMR device can further be enhanced via admixing of plasmonic signatures without much loss of the resolution.

Furthermore, these hybrid structures can set up a new paradigm in the field of strongly coupled systems leading to plasmonic hot electron generation and thus can find application in solar energy harvesting, photovoltaics, photocatalysis, and many others.

Methods:

Fabrication of Dielectric GMR Structure. TiO_2 films are deposited on glass substrates (2.5 × 2.5 cm2) using electron beam evaporation (LAB 500 evaporator, Leybold Optics GmbH) at a rate of 2 Å/s up to a layer thickness of 200 nm. They are divided into individual smaller pieces and spin-coated with the negative photoresist ma-N 405 diluted with ma-T 1050 (MicroChem) in a 1:1 ratio. Optimized spin parameters of 6000 rpm, acceleration of 2000 rpm/s, and a total spin time of 33 s produced a thin film of 150 nm thickness, as confirmed by spectroscopic ellipsometry (RC2-DI, J.A. Woollam Co., Inc.). The coated substrates are further exposed to LIL to obtain a photoresist grating.

Gold Particle Synthesis. Spherical gold NPs with 88 ± 12 nm diameter are synthesized via a colloidal seeded growth process according to the procedure of Bastus et al.[28] Citrate-stabilized NPs are coated with bovine serum albumin (BSA) prepared according to already published methods.[29] Briefly, 40 mL of the NP solution is added to 4 mL of 1 mg/mL BSA solution with 1 wt % sodium citrate at pH 9. The solution is allowed to incubate overnight. After fourfold centrifugation, the NPs are analyzed with UV–vis spectroscopy and transmission electron microscopy. The NP size of 88 nm is determined by a statistical analysis of 50 particles.

Directed Self-Assembly. The fabricated dielectric GMR sample is UV flood-exposed (UV-15 S/L, Herolab) at 254 nm for 10 min and subsequently hard-baked at 120 °C for 4 min on a hot plate to reduce the swelling and leaching out of the photoresist while in contact with the slightly alkaline particle solution. Oxygen plasma treatment (30 s, 0.2 mbar, 80 W, Flecto 10, Plasma Technology) prior to the convective assembly experiment[30] creates good wettability of the substrate that is fixed to a motorized translation stage (PLS-85, Physik Instrumente). A cleaned glass slide (Menzel) is mounted above the GMR sample at a distance of 0.5 mm, and 25 µL of NP solution (0.5 mg/mL gold, pH 9) is placed in between the gap. The elevated pH value ensures a strong negative surface charge of the protein coating (−30 mV)[29] to provide electrostatic stabilization of the colloidal suspension. A constant linear motion at the rate of 1 µm/s is imparted to the stage through a computerized software system to recede the contact line in a direction parallel to the channels. The stage temperature is set to 14 K above dew point to control the evaporation rate at the meniscus that drives the transport of NPs toward the three-phase contact line. Lateral confinement by the channel geometry and a vertical one determined by the thickness of the liquid film inside the channels enables the selective crystallization of particles into double (dimer) lines together with the help of attractive capillary forces that arise during drying.

UV–Vis–NIR Spectroscopy. An ultraviolet–visible–near-infrared (UV–vis–NIR) spectrophotometer (Cary 5000, Agilent Technologies) in transmission geometry is used for recording optical responses corresponding to a broad range of 400–2400 nm. A rotatable polarizer is used to investigate the effect of both polarizations, s-pol (transverse electric, TE) and p-pol (transverse magnetic, TM), where the electric fields are out-of-plane and in-plane to the plane of incidence, respectively.

Finite-Difference in Time-Domain Simulations. A commercial-grade simulator based on the finite-difference in time-domain (FDTD) method is used to perform the calculations (Lumerical FDTD,[31] version 8.16). For the simulation of the optical response, a plane wave source is used and the frequency points are set equal to that of the wavelength span. Monitor boxes (transmission monitors kept normal to the substrate) are used to obtain the optical responses of the systems. For the dielectric properties of gold, data from Palik[142] are fitted using six coefficients with a root-mean-square error of 0.2. For the photoresist, TiO_2 layer, and glass substrate, the optical constants are fitted by using the experimentally obtained values through spectroscopic ellipsometry that are available in Appendix 6.11 (see Figure 35). The mesh size in the FDTD region is set to auto-nonuniform with a minimum mesh size of 0.25 nm and an additional mesh overlay of 2 nm applied over the arranged particle geometry. Periodic boundary conditions are applied for X and Z directions with perfectly matching layers along the Y direction.

RI Sensing. Deuterium oxide of 99.9% purity (D_2O), commonly known as heavy water, is used as an initial surrounding medium which is altered with a solution (0.3 g/mL) of sodium chloride (NaCl) in D_2O to slightly increase RI and observe the characteristic spectral change. The RI of pure water and salt solution are initially measured using a digital multiple wavelength refractometer DSR-L (Schmidt + Haensch) and used for the calculation of the sensitivity of different resonant geometries.

2.3. Hot electron generation via guided hybrid modes

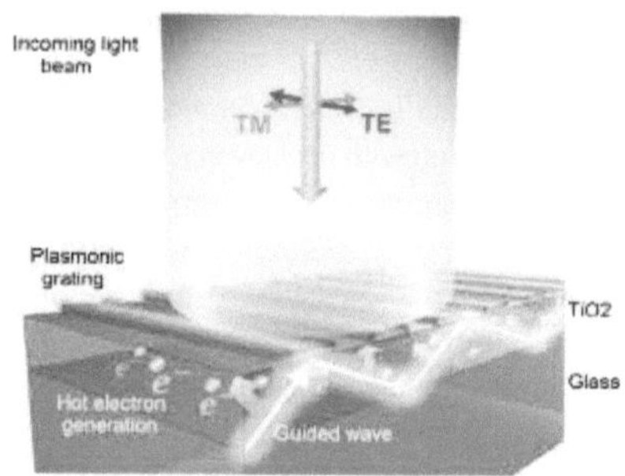

This chapter contains unpublished work, contributions to each part are to the date of the submission of the book.

By ‖ *Swagato Sarkar,[1,2]* ‖ ***Vaibhav Gupta,**[1] Jeetendra Gour,[1,2] Takuya Tsuda,[1] Olha Aftenieva[1], Arvind Singh[2], Sunil Kumar,[2] Anja Maria Steiner,[1,3] Joby Joseph,[2] Tobias A.F. König[1,3,]**

S. Sarkar, V. Gupta, J. Gour, O. Aftenieva, A. M. Steiner, Dr. T.A.F. König
Institute for Physical Chemistry and Polymer Physics, Leibniz-Institut für Polymerforschung Dresden e.V., Hohe Str. 6, 01069 Dresden, Germany.
S. Sarkar, J. Gour, A. Singh, Prof. S. Kumar, Prof. J. Joseph
Department of Physics, Indian Institute of Technology Delhi, New Delhi, 110016, India.
A. M. Steiner, Dr. T.A.F. König
Cluster of Excellence Center for Advancing Electronics Dresden (cfaed) and § Physical Chemistry of Polymeric Materials, Technische Universitat Dresden, 01069 Dresden, Germany.

‖ Equal contribution: S.S., V.G.

* Corresponding author

Author contribution statement

SS and **VG** have equally contributed to this work. **VG** encouraged and supervised JG and OA to fabricate the template and did the optical set up of two beam interference techniques. **VG** performed all assembly experiments. **VG**, SS and JG performed and analyzed all the spectroscopy measurements. SS performed all software-based (FDTD) simulations. TT performed all the photocurrent measurements. AS synthesized the gold nanoparticle. AS, SK and SS performed the pump probe experiments. JJ and TK contributed in developing the concept and writing of the manuscript. All authors provided critical feedback and helped shape the research, analysis and manuscript.

Introduction

The light matter interaction at nanoscale induces the collective oscillation of charge carriers in a metallic nanostructure. These conduction band electron has enormous energy densities at nanoscale which can be utilize for various photophysical processes.[143] "They are also known as "hot electron" because the energy they possess is much higher than room temperature thermal excitation.[116,144–149] The excitation of an electron in a metal results in two types of carrier: an electron and a hole.[150] Localized surface plasmons can decay radiatively via re-emitted photons or non-radiatively via excitation of hot electrons. In noble-metal nanostructures, non-radiative decay can occur through intraband excitations within the conduction band or through interband excitations resulting from transitions between other bands (for example, d bands) and the conduction band. Plasmonic energy conversion: electrons from occupied energy levels are excited above the Fermi energy and so these are called "hot electrons" further they can be injected into a semiconductor by forming a Schottky barrier with the plasmonic nanostructure. Hot electrons with energies high enough to overcome the Schottky barrier $\varphi_{SB} = \varphi_M - \chi_S$ are injected into the conduction band E_c of the neighboring semiconductor, where φ_M is the work function of the metal and χ_S is the electron affinity of the semiconductor.[150]

The process of excitation and relaxation can be understood with the illumination of a metal nanoparticle with a laser pulse, and characteristic timescales. First, the excitation of a localized surface plasmon redirects the flow of light (Poynting vector) towards and into the nanoparticle. In the first 1–100 fs following Landau damping, the athermal distribution of electron–hole pairs decays either through re-emission of photons or through carrier multiplication caused by electron–electron interactions. During this very short time interval τ_{nth}, the hot carrier distribution is highly non-thermal.[151] The hot carriers will redistribute their energy by electron–electron scattering processes on a **timescale** τ_{el} **ranging** from 100 fs to 1 ps. d, Finally, heat is transferred to the surroundings of **the metallic** structure on a longer time scale τ_{ph} ranging from 100 ps to 10 ns, via thermal conduction.[152]

The relaxation process for these higher energy plasmons **are via two** channels: radiative and nonradiative relaxation. In case of radiative relaxation **they re-emit a** photon and in the case of nonradiative relaxation it is decided by landau damping.[153] **After the** plasmon excitation at t=0 sec a very fast process (1-100 femto sec) of landau **damping occurs.** This is a non-thermal process where the electron hole pair decays either through **re-emission** of photons or through carrier multiplication due to electron interaction. Further **the hot carriers** will redistribute their energy by electron–electron scattering processes **such as Auger** transitions within 100 fs to 1 ps. In the end they can interact with the lattice phonons **and dissipate** in terms of local heat on the longer time scale of 100 ps to 10 ns. Therefore, **It is very** challenging to increase the yield of charge-carrier extraction from the plasmonic **structure. It is** also important to note that one should design the plasmonic structure in such **a way that it** could support a bigger optical cross section in terms of light absorption so that **it can induce more** electron hole pairs hence more photo current efficiency. However, when **it comes to** the designing of the structure waveguide plasmon polariton could support **localized resonances** and waveguide mode all together. These polaritons have contributions **from both** waveguide evanescent mode and plasmonic mode. The hybrid character **leads to the** suppression of radiative damping hence narrow line width. Gomez and co-**workers investigated** the photoinduced electron transfer in the strong coupling regime for Au **bar structures on** TiO2 film.[131]

We have made you familiar with the waveguide plasmon **polariton and** the advantages of exciting them into the system via rational design **parameters. It is well** known that when plasmon decay via nonradiative process they could create **so called hot** electron hole pair. If the metal is in direct contact with the suitable interface **such as semicond**uctor material these hot electron could result into more charge carriers into **conduction band.** Hot carrier transfer also depends on the shape and size of plasmonic.[154] **Further the quantum** yield to extract the hot electron is still low. Researchers are trying to impro**ve it via various** ways. Embedding the semiconductor material with metal to increase the **contact area seems** to be a fruitful teachnique.[155–157] However metamaterial perfect absor**ber could also be** a solution to the poor quantum efficiency. Li et al. has been reported near un**ity optical absorp**tion using ultrathin plasmonic nanostructures with thicknesses of 15 nm, s**maller than the** hot electron diffusion length. These perfect absorber based metamaterial **could serve as** a platform for the enhancing the efficiency of hot electron based photovol**taic, sensing and** photocatalysis.[158,159] Generally the photocatalysis process is not very **efficient**[160] but plasmon induced photocatalysis has come up as very effective approach.[161–163] **In this** way the hot electron transfer from the metal nanostructures transfer into photocatal**ytic mate**rial. This process can be understood as follows[164]: First step is the absorp**tion of photons** by metal so that the electrons in the metal can lift up from states below the **Fermi level by** the relevant photon energy E = hv where h is the plank constant. This **process can be** maximized under the condition where the metal absorbs more or less all inc**ident photons ac**ross a broad spectral bandwidth.[165] In the second step the excited electrons would try to cross the metal/ oxide barrier. The efficiency of this injection process is defined by the energy-momentum distribution of the hot electron population.[166] In a first order approximation one can assume the isotropic momentum distribution for the nanostructures. This assumption is true for semi-infinite planar surfaces and can not be extended for nanoparticles. Hot electrons are not free but tightly bound to the nanostructures so the conservation of momentum doesn't hold and more rigorous study is needed.[151] Finally few of the hot electrons would be injected to the junction without going to the inelastic collisions. The hot electrons coming to the junction with the energy exceeding the barrier φ_{SB} have a certain probability to cross the barrier but it can still be reflected. Zeng et al. discussed three different cases where they created three different junctions in nature Schottky, Ohmic and insulator.[131] For the case of glass (insulator) substrate, electron injection from Au bars is not possible due to unfavorable energetics. For an Ohmic case where the injection is possible but the lifetime of the charge separated states will be smaller in contrast to the Schottky contacts. In case of Schottky band bending, an energetic barrier prevents the injected electron from recombining with the holes in the metal and hence increases the charge seperation time. Strong coupling between plasmons and optical modes, such as waveguide or resonator modes, gives rise to a splitting in the plasmon absorption band. As a result, two new hybrid modes are formed that exhibit near-field enhancement effects. These hybrid modes have been exploited to improve light absorption in a number of systems. This modal strong coupling between a Fabry–Pérot nanocavity mode and a localized surface plasmon resonance (LSPR) facilitates water splitting reactions. We use a gold nanoparticle (Au-NP)/TiO2/Au-film structure as a photoanode. This structure exhibits modal strong coupling between the Fabry–Pérot nanocavity modes of the TiO2 thin film/Au film and LSPR of the Au NPs.[157,167]

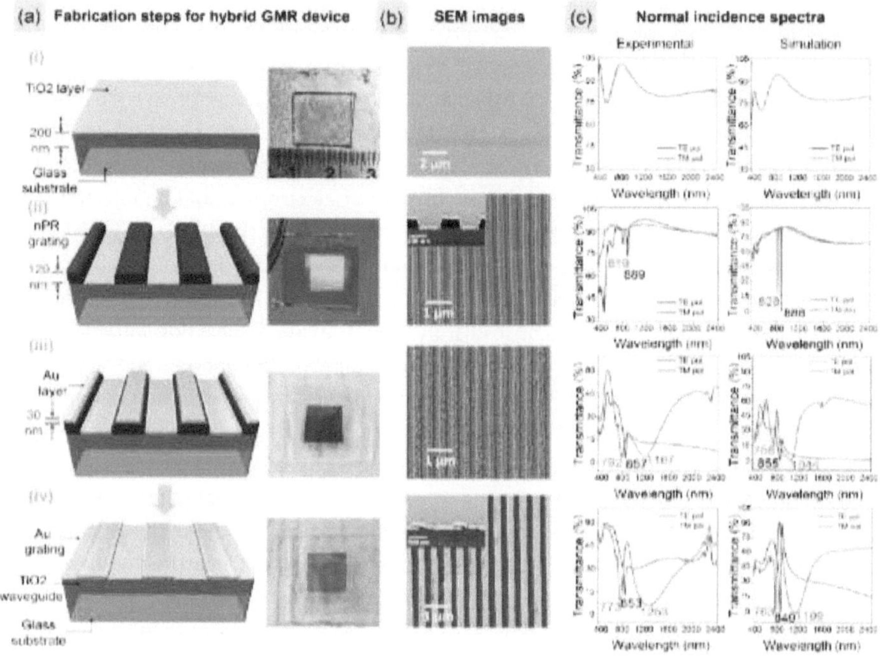

Figure 17. Fabrication and characterization of the hybrid opto-plasmonic structure. (a) The process steps for fabrication of the desired hybrid structure along with sample images are described in following steps: (i) 200 nm of TiO2 layer is deposited on glass substrate using electron beam physical vapor deposition. (ii) Negative photoresist (nPR) grating on TiO2 waveguide layer is formed by laser interference lithography (LIL) using a He-Cd laser. The resultant architecture is now a dielectric GMR structure with 500 nm period (iii) Deposition of 30 nm gold using thermal evaporation. (iv) Lift-off using an acetone bath via sonication process resulting in plasmonic grating on dielectric waveguide. (b) SEM image of the structures during fabrication in each of the process steps (i)-(iv). A FIB cut revealing the cross section is shown in the inset for the cases of (ii) photonic GMR and (iv) hybrid GMR. (c) Normal incidence spectra for both TE and TM cases obtained experimentally as well as through simulation is plotted for the cases (i) to (iv).

As we have discussed before, Gomez[131], Hiroaki[168] and coworkers investigated the photoinduced electron transfer in the strong coupling regime for Au bar structures on TiO_2 film. These experiments were performed on e-beam fabricated structures here we are taking it one step further in terms of large scale fabrication by introducing LIL fabricated structures on centimeter scale which can further be extended to meter scale.[4]

2.3.1. Fabrication of the hybrid GMR structure via LIL and lift-off process

A layer of 200 nm titanium dioxide was deposited on top of precleaned glass substrate as shown in figure 17. The deposition was done using e-beam evaporation and the layer serves the purpose of a slab waveguide. Further a negative photoresist mn-405 was coated on top of the TiO_2 layer and exposed to 325 nm laser in order to fabricate one-dimension grating. We have produced various periodicities ranging from 300 nm, 400 nm, 560 nm and make sure that there is no underneath photoresist via focused ion beam (FIB) cut. Further a 30 nm gold layer was deposited on top on the grating structures with the help of thermal deposition. Structures were soaked in acetone for 2 to 3 hours for smooth lift off. Further the sample was immersed in an acetone bath and lift off was done in combination with mild sonication. Finally, we had gold gratings on top of TiO_2 slab waveguide. All together it could also be called a guided mode resonance (GMR) device.[36] The whole fabrication procedure is depicted in figure 17a. Figure 17b shows the corresponding SEM images of such devices with FIB in the inset.

2.3.2. Spectroscopic and simulative analysis of hybrid opto-plasmonic structures of different periodicities

Later we optically characterize the device at each and every step of fabrication as shown in figure 17c. TiO_2 layers absorb most of the light in the UV region corresponding to the band gap of semiconductor material. The same has been confirmed via FDTD simulations as well. Further the photoresist grating on top of the slab waveguide acts as a GMR device as a whole. This leads to the sharp dip in the transmission spectra for both TE and TM cases as shown in figure 17c. We have discussed the case of 500 nm periodicity where GMR occurs at 888 nm and 828 nm for TE and TM cases respectively. These are pure photonic modes with narrow line width.

Finally, after gold deposition and lift off process dielectric GMR structure converts into metallic GMR device. Here it is important to discuss the origin of these modes. There is a plasmonic element called nanowire which exhibits localized surface plasmon resonance which depends on the width, height and period of the wire (in case of gratings). These plasmon resonances can be excited with the light perpendicular to the wire length. Under a specific incidence angle and for particular grating period, these nanowires grating can excite resonances involving coupling of energy into waveguide and hence gives arise the formation of waveguide-plasmon polariton.[169] These waveguide-plasmon polariton are the collective excitation of waveguide supported periodic plasmonic structures which are characterized by exhibiting sharp spectral linewidths that arise from the relative long lifetimes of the photonic character of these polaritons, in addition to having highly localized near-fields around the metal nanowires.[133,169,170]

In other words, the grating can contribute momentum to the incident light and couple it to the waveguide mode supported by TiO_2 slab waveguide. As we increase the grating period the GMR mode can be bought into resonance with the plasmonic nanowire resonance hence coupling occurs which leads to two hybridized states.[131] These spectral doublets give birth to a quasiparticle which is coherent superposition of plasmon and waveguide mode as shown in figure 17c.[130] We further performed FDTD simulations to quantify the same via solving Maxwell's equations and found the similar behavior. These hybrid modes show anti-crossing nature with the further progression of the grating **period which is** also a well-known characteristic of waveguide plasmon polariton.[167]

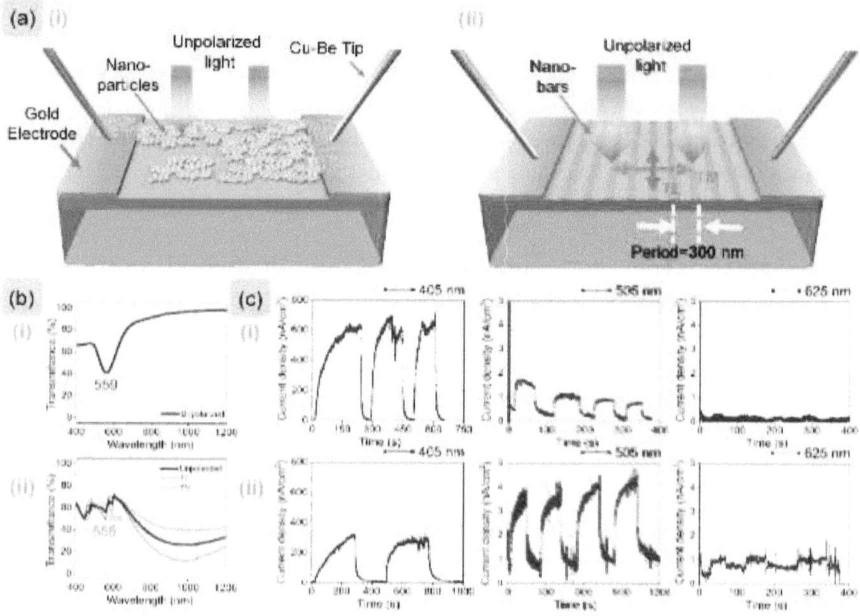

Figure 18. Photocurrent generation in different plasmonic nanostructures: (a) Schematic of two between two different geometries, (i) metallic nanoparticles **on TiO2 with particle** diameter around 88 nm and (ii) metallic nano-bars with period 300 nm on TiO2 **supporting the** desired hybrid mode. Unpolarized light at specific wavelengths is used throughout **this study. (b)** Resonant wavelengths are determined from the transmittance of the two different **systems: (i) particles** with resonant LSPR wavelength around 559 nm when excited with a broadband **unpolarized light** and (ii) the hybrid GMR, where unpolarized as well as polarized excitation reveals **556 nm as de**sired wavelength for hybridization of the TM component. (c) Photocurrent **generation data for** the case of periodic excitation for the two mentioned cases (i) and (ii) with three **different wavelengths** 405, 505 and 625 nm. 405 nm (~3.06 eV) gets mostly absorbed by the TiO2 **resulting in electron** hole pair generation, leading to the generation of photocurrent which is quite higher **than the other** cases. A 505 nm source can excite the particle resonance centered near 559 nm for **the case (i) where**as the TM component excites the hybrid mode in the case (ii) centered around 556 **nm. However, there** is a finite possibility that TiO2 absorption band is also affected. For 625 nm, being **far from the** resonance both the cases show almost no current generation in spite of periodic excita**tion.**

2.3.3. Comparative study of photocurrent generation in different plasmonic structures

After fabricating the GMR device we have performed the incident photon to charge carrier efficiency (IPCE) measurements. Rather than back we have employed the top electrode technique. The Au electrodes were deposited with physical vapor deposition via a mask. The mask was pre aligned in parallel to the grating direction before deposition with the mask aligner to avoid any cross talking in the later experiments. Apart from GMR devices we have also performed the control experiment by simply drop casting a AuNP solution on top of the TiO_2 waveguide. After drying, the optical spectrum confirms the presence of the spherical AuNP on top. It shows a LSPR at 559 nm for unpolarized excitation as shown in figure 18b(i). Similar spectra have been recorded for the metallic GMR device. It exhibits two hybrid modes for the case of TM (perpendicular to the nanowire long axis) and polarization. It doesn't show any waveguide plasmon polariton characteristic in case of TE polarization as shown in figure 18b(ii). Now for further experiments we will only consider the TM polarization. For the drop casted sample, we excited with three different high-power light emitting diodes (LEDs) 405, 505 and 625 nm. In case of 405 nm mostly excited the TiO_2 but in case of 505 nm wavelength entered inside the LSPR bandwidth and the photocurrent generation was due to the hot electron excitation. The reason for the low current values could be the less detection area between two electrodes. During the measurement we have applied a constant voltage of 5volt between the electrodes. As we moved further away from the spectral bandwidth of the LSPR and excited the sample with 625 nm which lies outside the LSPR spectrum we observed negligible amount of photocurrent as can be seen in figure 18c(i).

Similar experiments were further repeated for the metallic (Au) nano bar based GMR device which resembles the same behavior as the drop casted sample. There was a noticeable difference for the case of 505 nm excitation that the amount of current is two times higher in this case as compared to the drop casted sample as shown in figure 18c(ii). This could be explained with the previously discussed theory of waveguide-plasmon polariton. The wavelength 505 nm lies inside the spectrum of the hybrid mode which is partly excited during unpolarized incident light. This induces the quasiparticle or so called waveguide-plasmon polariton which has a longer lifetime and increase the interaction time between waveguide and enhanced e-field plasmonic element.

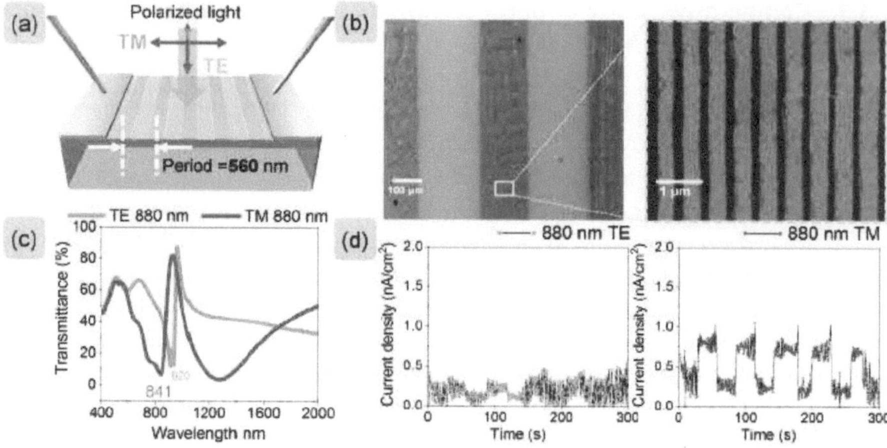

Figure 19. Polarization dependent response at higher wavelength: (a) Schematic of the desired configuration for the hybrid GMR with period 560 nm to match with available source of 880 nm. At a wavelength regime far away from the absorption band of TiO2, this strengthens the fact that photocurrent can still be generated without any influence of TiO2 electron-hole pair generation which may have been the case for exciting 300 nm period hybrid GMR with 505 nm (2.46 eV) as discussed in Figure 4. (b) An SEM image of the fabricated sample over a large area showing deposited gold electrodes for the photocurrent measurement. A magnified view shows the metallic bars within the electrode with a period of 560 nm. (c) The fabricated sample is inspected under a spectroscope to reveal the hybridized GMR mode at 841 nm. (d) Both TE and TM modes are separately excited using a polarizer. As seen, the TM mode responds to the periodic excitation whereas for the TE case the data is noisy. Thus, one can confirm that the hot electron generation is an outcome of the supported hybrid mode present only for the TM excitation case.

2.3.4. Polarization dependent response at higher wavelength

Further for the case of 625 nm no such current is observed because it lies out of the hybrid mode spectrum. In order to understand the hybrid nature of the modes (waveguide plasmon polariton) we fabricated GMR devices with various periodicities but initially the photocurrent measurement was performed on 300 nm periodicity samples. Later in order to strengthen the fact that photocurrent is generated without any influence of TiO2 electron-hole pair generation which may have been exciting in the case of 300 nm periodicity grating. Here we consider a bigger periodicity of 560 nm where there is no chance of exciting TiO$_2$ as shown in figure 19b. However previously we have made a claim that in case of metal bar polariton can be excited only under TM polarization.[169,171] To prove this analogy we have first taken the optical spectrum for the cases TE and TM. TM is excited at 841 nm and TE is excited at 920 nm as can be seen in the figure 19c. Further photocurrent was measured with the similar top gold electrode technique with 880 nm laser excitation. We have found that only for the case of TM we could observe the photocurrent but for TE the signal was very poor in quality. Even though

880 nm laser bandwidth overlapped with both the modes but current was observed only for TM cases.

Further In our previous publication Sarkar et al. has proved that similar waveguide-plasmon polariton can be excited in the nanoparticle chains with TE polarized excitation.[36] Further we would like to extend this realization in terms of photocurrent generation with the colloidal building blocks. In order to do that we have fabricated the hybrid structure of dimer chain plasmonic grating with the help of LIL and template assisted self-assembly.

(A) (B)

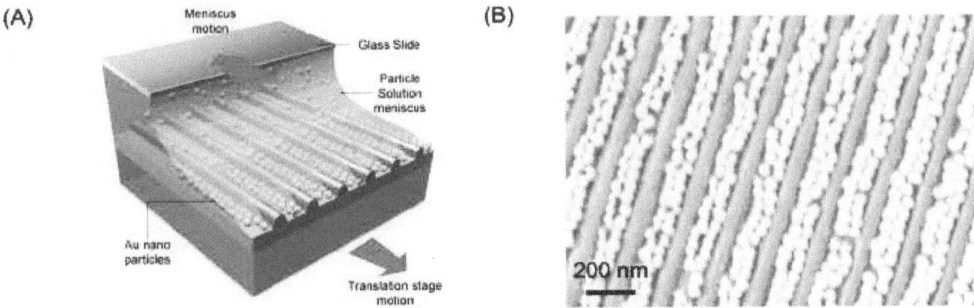

Figure 20. Directed self-assembly of gold nanoparticles within grating channels of a dielectric GMR structure supported by titanium dioxide film. (a) Schematic of the capillary assisted self-assembly. (b) SEM image of dimer nanoparticle chains.

2.3.5. Directed self-assembly of gold nanoparticles within grating channels of a dielectric GMR structure supported by titanium dioxide film

For the hybrid structure, synthesized plasmonic gold nanoparticles are required to be incorporated within the channels of the fabricated dielectric GMR structure. Due to spherical symmetry, concern about the final orientation of these nanoparticles relative to the channel does not arise. In the convective assembly experiment, 25 µL of gold nanoparticle solution (0.5 mg/mL, pH 9) is confined between the hydrophilized (30 s O_2-plasma, 0.2 mbar, 80 W, Flecto 10, Plasma Technology) topographical nano-channel substrate and a stationary glass slide (distance 0.5 mm). The substrate is withdrawn at a speed of 1 µm/sec to make the meniscus recede in the direction parallel to the channels. The evaporation, facilitated at the meniscus, drives a convective flow of particles towards the contact line. On the well wetting substrate, the fluid spreads inside the channels due to the capillary pressure. The nanoparticles travel along the liquid-filled channels that create both lateral and vertical (liquid film thickness) confinement. As soon as the particles protrude from the liquid film upon drying, attractive capillary forces arise between the particles to guide the crystallization into dimer lines of particles (Figure 20B).

Conclusion

In conclusion our previous publication Sarkar et al. has proved that similar waveguide-plasmon polariton can be excited in the nanoparticle chains with TE polarized excitation.[36] Further we would like to extend this realization in terms of photocurrent generation with the colloidal building blocks. Later pump and probe methods would be employed to understand the coupling regime in both cases: metal bars with TM and metal nanoparticle chains with TE polarization. Ultrafast spectroscopy could reveal that the electron resides inside the semiconductor material. We have created a similar GMR device with nanoparticle chains in our earlier publication where template assisted self-assembly was used as a tool to fill the photoresist grating channels which was created via LIL on a large scale. Hence, we present an alternative to the e-beam fabricated structures together with the alteration of polarization with simplicity of colloidal way for cost-effective and large scale plasmon assisted photocurrent enhancement. We hope that this might pave the roadmap for better photocatalysis, photodetection and sensing activities.

Methods

Fabrication of dielectric GMR structure: TiO2 films are deposited on glass substrates (2.5 x 2.5 cm2) using electron beam evaporation (LAB 500 evaporator, Leybold Optics GmbH) at 2 A°/s rate up to a layer thickness of 200 nm. They are divided into individual smaller pieces and spin coated with negative photoresist ma-N 405 diluted with ma-T 1050 (MicroChem) in 1:1 ratio. Optimized spin parameters of 6000 rpm, 2000 rpm/s acceleration and a total spin time of 33 s produced a thin film of 150 nm thickness as confirmed by spectroscopic ellipsometry (RC2-DI, J.A. Woollam Co., Inc.). The coated substrates are further exposed to laser interference lithography to obtain photoresist grating. Once fabricated, they are processed through plasma etch (system name with conditions) for 4-5 minutes to ensure grating depth upto TiO2 film surface.

Lift-off techniques towards hybrid structure: Once the dielectric GMR structure with photoresist grating and TiO2 waveguide is fabricated and characterized, it is further coated with 30 nm of gold film through thermal evaporation. Due to complete removal of photoresist from grating channels, gold is deposited in the form of bars within the channel directly onto TiO2 surface as well as on the top of the grating ridges. For the sonication process, complete immersion of the gold deposited samples in acetone is carried out in an ultrasonic bath at 80 Hz, with power starting from 30 Watt till 100 Watt.

UV-Vis-NIR spectroscopy: UV-Vis-NIR Spectrophotometer (Cary 5000, Agilent Technologies) in transmission geometry is used for recording optical responses corresponding to a broad range of 400 to 2400 nm. A rotatable polarizer is used in order to investigate the effect of both the polarizations, s-pol (transverse electric, TE) or p-pol (transverse magnetic, TM) where the electric fields are out of the plane and in-plane to the plane of incidence, respectively.

Finite-difference in time-domain (FDTD) simulations: A commercial-grade simulator based on the finite-difference in time-domain method is used to perform the calculations (Lumerical FDTD, version 8.16). For the simulation of the optical response, a plane wave source is used and the frequency points are set equal to that of the wavelength span. Monitor boxes (transmission monitors kept normally to the substrate) are used to obtain the optical responses of the systems. For the dielectric properties of gold, data from Palik is fitted using six coefficients with an RMS error of 0.2. For the photoresist, The mesh size in the FDTD region is set to auto non-uniform with a minimum mesh size of 0.25 nm with an additional mesh overlay of 2 nm applied over the arranged particle geometry. Periodic boundary conditions are applied for X and Z directions with perfectly matching layers along the Y direction.

Directed Self-Assembly: The fabricated dielectric GMR sample is UV flood-exposed (UV-15 S/L, Herolab) at 254 nm for 10 min and subsequently hard-baked at 120 °C for 4 min on a hot plate to reduce the swelling and leaching out of the photoresist while in contact with the slightly alkaline particle solution. Oxygen plasma treatment (30 s, 0.2 mbar, 80 W, Flecto 10, Plasma Technology) prior to the convective assembly experiment30 creates good wettability of the substrate that is fixed to a motorized translation stage (PLS-85, Physik Instrumente). A cleaned glass slide (Menzel) is mounted above the GMR sample at a distance of 0.5 mm, and 25 µL of NP solution (0.5 mg/mL gold, pH 9) is placed in between the gap. The elevated pH value ensures a strong negative surface charge of the protein coating (−30 mV)29 to provide electrostatic stabilization of the colloidal suspension. A constant linear motion at the rate of 1 µm/s is imparted to the stage through a computerized software system to recede the contact line in a direction parallel to the channels. The stage temperature is set to 14 K above dew point to control the evaporation rate at the meniscus that drives the transport of NPs toward the three-phase contact line. Lateral confinement by the channel geometry and a vertical one determined by the thickness of the liquid film inside the channels enables the selective crystallization of particles into double (dimer) lines together with the help of attractive capillary forces that arise during drying.

Photocurrent measurement: Gold contacts (30 nm thick, channel length: 150 µm, width: 2.2 mm) were deposited on the metal gratings by thermal evaporation in vacuum through a shadow mask. The photocurrent measurements were carried out by using the two-point probe method. Keithley 2612B was used to measure the I-V characteristics and the applied bias voltage was 10V. The light emitting diode (LED) (Thorlabs, LED4D140, wavelength: 405, 505, 590, 625 nm and M880L with the collimation adapter, wavelength: 880 nm) was used as a light source. The polarizer (Thorlabs) was placed between the substrates and the LED light source. All the measurements were made under ambient conditions.

2.4. Active Chiral Plasmonics Based on Geometrical Reconfiguration

This chapter contains unpublished work, contributions to each part are to the date of the submission of the book.

By ∥ *Patrick T. Probst[1]*, ∥ *Martin Mayer[1,2]*, **Vaibhav Gupta[1]**, *Anja Maria Steiner[1,2]*, *Günter Auernhammer,[1]* *Tobias A. F. König[1,2,**], *Andreas Fery[1,2,**]*

P.T.Probst, M. Mayer, V. Gupta, A. M. Steiner, G. Auernhammer, Dr. T.A.F. König, A. Fery.
Institute for Physical Chemistry and Polymer Physics, Leibniz-Institut für Polymerforschung Dresden e.V., Hohe Str. 6, 01069 Dresden, Germany.
M. Mayer, A. M. Steiner, Dr. T.A.F. König, A. Fery
Cluster of Excellence Center for Advancing Electronics Dresden (cfaed) and § Physical Chemistry of Polymeric Materials, Technische Universität Dresden, 01069 Dresden, Germany.

∥ Equal contribution: P.T.P., M.M.

* Corresponding author

Author contribution statement

PP and MM have contributed equally to this work. PP performed the assembly experiments and did the SEM, UV-vis-NIR, AFM, CD spectroscopy characterizations. PP has carried out the 2-dimension finite element analysis simulation of strain induced bending. MM supported the advanced ellipsometry, software-based simulation (FDTD). **VG** fabricated the template and assisted with the assembly experiments as well as synthesized the required nanoparticle for the project. GA has provided his unprecedented guidance and devoted his time for scientific discussions. TK and AF have developed the concept, provided their guidance and feedback and were involved in the review and scrutiny process of the manuscript. All the authors have provided critical feedback and helped shape the research, analysis and manuscript.

Introduction

Light-induced excitation of collective electron oscillations also known as localized surface plasmon resonance in metal nanostructures focuses high electric fields to regions beyond the diffraction limit. Tuning these resonances in an in-situ manner in so-called active plasmonic systems[21] allows realization of colorimetric sensors,[172] tunable colour[84] and polarization filters[173,174], optical switches[175] and modulators.[176] Such systems employ the tunable nature of responsive polymers[99], phase-change materials[177], liquid crystals[178], magnetic particles[179], DNA hybridization[180] and elastomeric substrates.[4] The induced change in dielectric environment, charge carrier density or geometrical configuration leads to a shift in the resonance position. Amongst the different strategies, mechanical strain-induced modification of inter-particle distances proved to offer a simple and reversible real-time control over peak position and mode structure of the collective plasmon resonances.[106,181,182] Mechano-optical continuous tuning was recently exploited for matching nanolasing conditions and opens possibility as strain sensors.[85]

Current efforts are made to introduce chirality to the well-established schemes for dynamic optical control to gain enantioselective molecular detection[183] and control over circular polarizations. Chirality arises from the absence of a mirror and inversion symmetry in the nanostructure. Consequently, two enantiomorphs can exist as non-superimposable mirror images to each other and are characterized by their inverted chiroptical response. In particular, they can show inverse signs of optical rotation (OR) for linearly polarized light and inverted circular dichroism (CD), which defines the differential absorbance for left (LCP) and right circularly polarized light (RCP).

Although continuous modulation of CD magnitude and its spectral tuning can be easily achieved using well-established strategies from the field of active plasmonics, the inversion of chirality remains a key challenge. 3D geometrical conversion between two enantiomers has been achieved in solution-based DNA hybridization.[184–186] Since these are usually restricted to diluted systems, however, the achieved chiroptical effects remain small. To overcome this limitation, dense and switchable plasmonic arrays would be desirable to achieve substantial modulation of circular polarizations. Microelectromechanical systems (MEMS)[187,188] and an alternative design using a bias layer to avoid the need for geometrical reconfiguration[189] were shown to switch between two states of enantiomorphs in an on-off manner, only at the cost of elaborate fabrication schemes. Furthermore, gradual conversion from one enantiomorph to the other one remains elusive, especially if strong chiroptical effects are desired.[190] Systems that exploit the full potential of chirality tuning, i.e. gradual modulation of CD in sign, magnitude and spectral position simultaneously are still missing. Design limitations and the general challenges of 3D fabrication represent the main bottleneck. This becomes even more complicated when in combination with elastic substrates.[191,192]

2.4.1. Chiral 3D assemblies by macroscopic stacking of achiral chain substrates

Here, we build on our experience in the bottom-up fabrication of mechano-tunable 2D plasmonic lattices[4] to realize layered 3D reconfigurable chiral arrays with pronounced chiroptical effects. Approaching two achiral nanoparticle chains assembled in compliant nanochannels to nanometre distance yields 3D intrinsic chirality.[193] Our design uses the nanochannel walls both as a template for orientational control and as an elastic spacer for planarization of the two particle layers to realize small inter-layer distances. We demonstrate continuous 3D geometrical reconfiguration by reversible re-stacking at varying inter-layer rotations and by applying compressive strain normal to the interface. These two handles enable tuning the CD response in magnitude, sign as well as spectral position. Mechanical and electromagnetic simulations suggest that the particle chains bend upon compression, which results in a significant spectral shift. Based on the strong chiroptical effects and simple fabrication route, the particle based chiral design holds great potential as mechano-tunable circular polarization filter and strain sensors. In particular, the particle-based superstructures provide large volumes of super-chiral fields that would facilitate low detection limit chiral sensing of analyte solutions sandwiched between the two particle layers.[194]

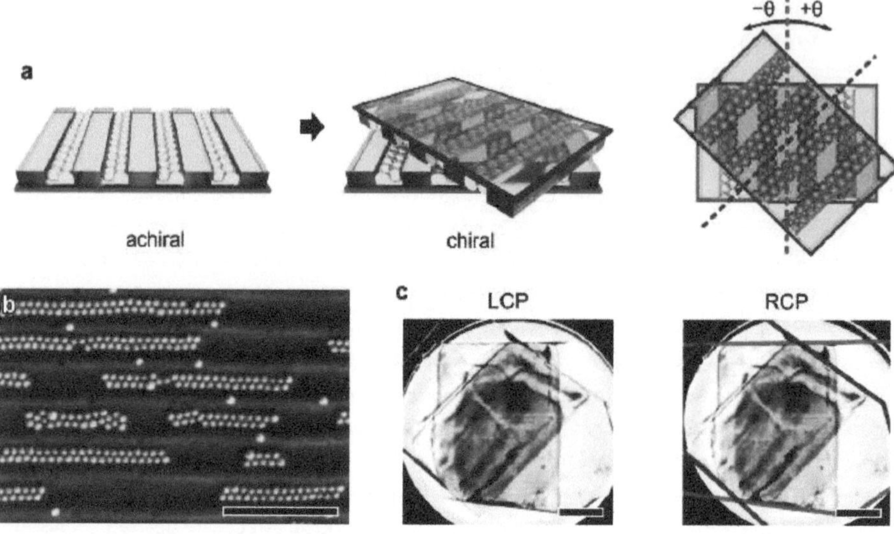

Figure 21: Chiral 3D assemblies by macroscopic stacking of achiral chain substrates. a, 3D scheme of stacking approach. The top view (right panel) illustrates the angle ⍰, which the bottom (red) and top sample (blue) are stacked at. **b,** SEM image of gold nanoparticle chains assembled inside PDMS nanochannels. Scale bar, 1 μm. **c,** Photographs demonstrating the polarization-sensitive transmission of a sample stacked −45° when illuminated with LCP and RCP light. Lower and upper samples are indicated in red and blue respectively. Scale bar, 2 mm.

We used 77 ± 2 nm gold nanoparticles as monodisperse colloidal building blocks to construct 3D chiral nanostructures. In a first capillarity-assisted particle-assembly step, the nanoparticles were assembled into nanochannels that were fabricated via laser interference lithography (Fig.21a). To this end, a droplet of nanoparticle suspension was dragged perpendicular to the channel features across the topographical template by a motorized movement.[107] For further experimental details please refer to the Methods section. The geometrical confinement within the channels and inter-particle attractive immersion capillary forces that arise upon drying produced double chains of gold nanoparticles (Fig. 21b). The anisotropic arrangement of particles into achiral chains over macroscopic (~1 cm²) areas renders the individual samples polarization-controlled plasmonic colour filters. The dichroic ratio of 0.81 for the longitudinal mode at 1048 nm demonstrates the high yield in unidirectional particle chains.

Macroscopic stacking (at oblique angles) of two such achiral particle chains supported on PDMS substrates resulted in a chiral 3D arrangement of particles as depicted in Fig. 21a. The vast possibilities of chiral configurations were characterized by the inter-layer rotation angles (θ) between the lower (red border) and upper substrate (blue border). Note that two enantiomorphs share the same magnitude but opposite sign of θ. The chirality of the 3D particle assembly corresponded to a pronounced CD on macroscopic scales that is visible to the bare eye. In the photographs in Fig. 21c taken under polarized illumination, the double layer of chains stacked at $\theta = -45°$ absorbs less in the magenta fraction of LCP light as compared to RCP light. In contrast, the individual achiral monolayers did not interact specifically with either of the two orthogonal circular polarizations.

Conclusion

In conclusion, as demonstrated, simple designs cost-effectively fabricated via bottom-up techniques even stand the comparison with e-beam structures for nanoplasmonic devices. Additional tunability of the strong optical effects by external stimuli facilitates broad-band applications and opens the door for sensing applications. Two achiral nanoparticle chain arrays on compliant substrates are placed on top of each other to produce a chiral 3D arrangement of rotated chains. To this end, convective assembly[75] of gold nanoparticles into nanochannels was employed. In this colloidal self-assembly process, a droplet of nanoparticle suspension is dragged over the topographical template in direction perpendicular to the channels. The evaporation of the fluid facilitated at the receding meniscus drives a particle flow toward the three-phase contact line.[76,77,107,127,195–200] Geometrical confinement exerted both by the channel geometry as well as the thin liquid film height guide the assembly into particle lines. The resulting small inter-particle and inter-layer distances allow for pronounced chiral plasmonic coupling.[201]

Methods

Nanoparticle Synthesis: Spherical gold nanoparticles were synthesized in a seed-mediated growth process[106], followed by PEG coating. Briefly, tetra chloroauric acid (HAuCl4) was reduced using sodium borohydride (NaBH4) in the presence of hexadecyltrimethylammonium bromide (CtaB) as stabilizing agent. The as-prepared 2 nm large, single-crystalline seeds were successively overgrown with additional gold to reach their final size of 77 ± 2 nm (TEM statistics 260 particles). The growth solution contained $HAuCl_4$, ascorbic acid, and CtaC as Au precursor, reductant and stabilizing agent, respectively. For the last growth step, a syringe pump setup was used to assure kinetic control over particle shape. After purification by centrifugation and washing, the surfactant concentration was set to 2 mM. Finally, the ligand (CtaC) was exchanged against PEG-6k-SH.[202]

Template-assisted colloidal self-assembly: The aqueous assembly solution contained PEGylated gold nanoparticles equal to a gold concentration of 0.5-1 mg mL^{-1} and a commercial trisiloxane surfactant (CoatOSil 77, Momentive). The surfactant lowers the surface tension significantly to ~21 mN m^{-1} and improved wetting of the nanochannel template. The low surface tension is crucial for successful capillarity-assisted particle assembly of nanoparticles.[107] A droplet of this nanoparticle solution was confined between the PDMS nanochannel template and a microscope slide (Menzel) using a custom-built setup. The glass slide used was hydrophobized with (Heptadecafluoro-1,1,2,2-tetrahydrodecyl) dimethylchlorosilane) (95 %, abcr) via gas-phase prior to assembly and adjust to a distance of ~500 μm above the PDMS substrate. Motorized movement of the sample (stepper motor PLS-85, Physik Instrumente) at speeds of 1-5 μm s^{-1} led to a continuous movement of the liquid meniscus across the substrate in direction perpendicular to the nanochannels. The assembly experiments were carried out ~11 K above the dew point.

Polarized Photography: The samples were illuminated from the back by polarized white light and the transmitted light was captured with a conventional compact camera. To create circular polarizations (LRC/RCP), the white light passed successively a linear polarization (Nikon, also used for linearly polarized photographs) and a quarter-wave plate (B. Halle, 460-680 nm).

UV-Vis-NIR spectroscopy: The linearly polarized spectra were measured with a Cary 5000 spectrophotometer (Agilent, USA) equipped with the Cary universal measurement accessory. The measurements were carried out in transmission under normal incidence of the polarized light and the spot size was 3 x 4 mm². The linear dichroic ratio was calculated as follows:

$$DR = \frac{E_\parallel - E_\perp}{E_\parallel + E_\perp}$$

where $E_\parallel$ and $E_\perp$ represent the extinction measured for linear polarization of the incident light parallel and perpendicular to the nanoparticle chain orientation, respectively.

Circular dichroism measurements: The circular dichroism was measured by Müller matrix (MM) transmission ellipsometry. Transmission ellipsometry was performed in the wavelength range from 193 nm to 1690 nm using a RC2 spectroscopic ellipsometer (J.A. Woollam Co., USA) equipped with focusing lenses, which allows direct measurement of all 16 MM elements. As discussed by Kumacheva and co-workers in detail, the detected intensity for LCP/RCP is defined by the MM interacting with its Stokes vectors as follows:[202]

$$\vec{S}_{out} = \hat{M} \cdot \vec{S}_{in}$$

where $\vec{S}_{in} = \begin{pmatrix} S_0 \\ S_1 \\ S_2 \\ S_3 \end{pmatrix}$ corresponds to $\begin{pmatrix} 1 \\ 0 \\ 0 \\ -1 \end{pmatrix} / \begin{pmatrix} 1 \\ 0 \\ 0 \\ 1 \end{pmatrix} / \begin{pmatrix} 1 \\ 0 \\ 0 \\ 0 \end{pmatrix}$ for LCP / RCP/ unpolarized excitation.

This results in the simple equations

$$I_{LCP} = I_0(\hat{M}_{11} - \hat{M}_{14})$$

$$I_{RCP} = I_0(\hat{M}_{11} + \hat{M}_{14})$$

$$I_{unpol} = I_0(\hat{M}_{11})$$

with I_0 being the transmitted light intensity of a stack of empty PDMS templates. The extinction spectra of LCP, RCP, and unpolarized excitation are plotted.

Thus, the extinction difference between LCP and RCP is defined by $\Delta Ext = -\log\left(\frac{1-\hat{M}_{14}}{1+\hat{M}_{14}}\right)$ allowing to calculate the circular dichroism according to $\theta = \left(\frac{\ln 10}{4}\right)\left(\frac{180}{\pi}\right)\Delta Ext$.

Atomic Force Microscopy (AFM): AFM height images were measured using a Dimension 3100 NanoScope V (Bruker) operated in tapping mode. Stiff cantilevers (40 N m-1, 300 kHz, Tap300, Budget Sensors, Bulgaria) were employed.

Scanning Electron microscopy (SEM): A NEON 40 FIB-SEM workstation (Carl Zeiss Microscopy GmbH) operated at an accelerating voltage of 1 kV was used for capturing scanning electron micrographs.

Electromagnetic Simulations: For the electromagnetic simulations, a commercial-grade simulator based on the finite-difference time-domain (FDTD) method was used to perform the calculations (Lumerical Inc., Canada, version 8.16). Circular polarized light was achieved by the superposition of the complex electric and magnetic fields of two separate simulations. The sources of these simulations feature orthogonal linear polarizations and a phase difference of $\pm 90°$ in order to obtain left/right circular polarized light, respectively. For the dielectric properties of silver, the data from Johnson and Christy was used. The crossed particle lines were modelled with a particle size of 70 nm, an interparticle distance (within one chain) of 2 nm, and an inter–chain distance of 5 nm. The PDMS embedding was incorporated into the model by an effective surrounding medium with a refractive index of 1.45. Simulation space was meshed with 1 nm and was surrounded by perfect absorbing boundary conditions (PML). To determine the field distributions, the model was simulated at the wavelengths of the corresponding plasmonic (chiral) modes. All simulations reached a convergence of 10^{-6} before reaching 500 fs simulation time.

3. Conclusion

Contemporary advances in the field of plasmonics are closely linked with the evolution of and developments in rational design and fabrication of plasmonic nanostructures. Within this context, my book focused on the large-scale, template-assisted self-assembly of well-defined, isotropic metal nanoparticles to generate tailored plasmonic surface-assemblies with strong light matter coupling and novel optical properties.[203–205] We aimed toward the admixture of top-down (LIL) and bottom up (TASA) approaches to produce macroscopic structures with nanometer precision (e.g. highly selective orientation and small gap sizes). Periodicities spanned from 200 nm to 1 micron. These assemblies resulted into the particular system order which gives rise to optical effects such as Bragg diffraction or GMR. When individual plasmonic building blocks[206] interact with these optical effects they further generate the collective optical properties such as SLR, PGR, chiro-optical effects, photo-current enhancement via hot-electron injection and many more. These collective effects hold great potential for various applications, *e.g.* plasmon lasing, high quality plasmon sensing, surface-enhanced Raman scattering (SERS), optical metamaterials etc.[127,207,208]

In my book, I mainly focused on TASA of spherical gold nanoparticles organized on PDMS templates in 2D fashion. We focused on two differently-coated particle types; i.e. the Au_BSA system and Au_PEG system. Finally, we achieved SLR response in these 2D plasmonic lattices which was tunable in real time due to the mechanical flexibility of the template. I also discussed the assembly of ordered nanoparticle arrays that show a narrow and mechanically tunable SLR over 100 deformation cycles of deformations. The assembly relies on directed self-assembly of colloidal nanoparticles in templates formed by soft lithography. We demonstrated the arrangement with both broad/ narrow particle size distributions and discussed their influence on the optical quality. The real-time mechanical tuning and accessibility of non-degenerated SLR modes hold great potential for plasmonic lasing,[110] biosensing,[209] or colorimetric sensors.[210] We also showed that the optical properties of the system change dramatically upon application of strain, as the array changes from a square lattice to a rectangular lattice. Importantly, the optical quality is not lost upon straining the system and the SLR peaks can be shifted over a wavelength range of 70 nm. Moreover, the uniaxially strained lattice opens up the possibility for introducing optical anisotropy **(Chapter 2.1).**

Further in this direction I established TASA of BSA-coated AuNPs in a 1D lattice geometry. In particular, the assembly was done directly on a photoresist grating which reduces the number of steps involved in fabrication and eases the fabrication process. We have established this process on TiO_2 (200 nm thick film) and gold (35 nm) films which were supported by a glass substrate. This leads to hybridized guided modes such as PGR and out of plane lattice resonances. The fabricated structure with a proper choice of design parameters supports coupling of the plasmonic radiant modes of the gold NP grating to t based GMR under normal incidence; this results in the formation of hybridized states that can be tuned by varying AOI. The hybridized structure was optically characterized in comparison to its constituent resonant geometries along with an in-depth study of these resonances through numerical simulations to understand the coupling of two resonances. On the basis of the electric field plots, a matrix transfer model is also established that could be more generally applied in general to similar cases of coupling of different resonances. With a completely new system in the field of modal strong coupling, we have realized its potential application in RI sensing where the sensitivity of a GMR device can further be further enhanced via admixing of plasmonic signatures without much loss of the resolution **(Chapter 2.2)**.[211]

In our above-mentioned study, we have proven that waveguide-plasmon polaritons can be excited in the nanoparticle chains with TE polarized excitation. To extend this concept we generated photocurrent with the colloidal building blocks. The pump/probe method was then employed to understand the coupling regime in both cases: metal bars with TM polarization, and metal nanoparticle chains with TE polarization. Ultrafast spectroscopy revealed that the electron resides inside the semiconductor material. We also created a GMR device with nanoparticle chains where template assisted self-assembly was used as a tool to fill the photoresist grating channels fabricated via LIL on a large scale. Hence, we were able to present an alternative to the e-beam fabricated structures together with the alteration of polarization with cost-effective and large scale plasmon assisted photocurrent enhancement. We hope that this might act as a roadmap toward better photocatalysis, photodetection and sensing activities **(Chapter 2.3)**.[212]

Further we stacked two substrates with particle chain arrays, and in doing so we created a simple yet powerful strategy to reversibly control all three CD characteristics (magnitude, sign and spectral position). The colloidal approach benefits from cost-efficient fabrication, while generating pronounced chiroptical response. The extended superchiral fields that arise in the vicinity of the chiral nanostructure can boost sensitivity in enantio-selective molecular detection. To this end, analytes can be easily applied in the cavity of the particle stack. Moreover, the strongest CD modes appear in the first (650-950 nm) and second (1000-1700 nm) transparency window for biological tissue[31]. Consequently, the introduced system is well suited for analysing biological samples. Real-time tunability of CD by re-stacking or compression fosters the development of compact spectroscopic devices and light modulators. As a unique feature, the introduced system enables strain-induced local tunability of chiro-plasmonic properties. The mechanistic understanding of this phenomenon by means of mechanical and electromagnetic simulations provides the basis for generalization of the observed effect and rational design of chiroptical properties. In particular, this mechanism serves as a new paradigm for design of active chiral devices, highly integrated sensors and complex structuring of light **(Chapter 2.4)**.[213,214]

Finally, the amalgam of top-down and bottom-up approaches developed in this context allows assembly of nanoparticles with different geometries and compositions (for *e.g.* core-shell particles, cube, cuboid, rod-like particles of Au, Ag). Finally, the rational design framework we established also allows us to extend the approach to supracolloidal structures (clusters, hetero-clusters, etc.).[99,213,215,216]

4. Zusammenfassung

Die gegenwärtigen Fortschritte auf dem Gebiet der Plasmonik sind eng mit der Evolution und den Entwicklungen beim rationalen Design und der Herstellung plasmonischer Nanostrukturen verbunden. In diesem Zusammenhang konzentrierte sich meine Arbeit auf die großflächige, templat gestützte Selbstorganisation gut definierter isotroper Metallnanopartikel, um maßgeschneiderte plasmonische Oberflächen Anordnungen mit starker Kopplung der leichten Materie und neuartigen optischen Eigenschaften zu erzeugen.[203–205] Wir haben uns zum Ziel gesetzt, Top-Down- (LIL) und Bottom-Up- (TASA) Ansätze zu mischen, um makroskopische Strukturen mit Nanometer Genauigkeit (z. B. hochselektive Orientierung und kleine Spaltgrößen) und Periodizitäten von 200 nm bis 1 Mikron herzustellen. Baugruppen führen zu einer bestimmten Reihenfolge im System, die optische Effekte wie Bragg-Beugung oder GMR hervorruft. Wenn einzelne plasmonische Bausteine[206] mit diesen optischen Effekten interagieren, erzeugen sie ferner die kollektiven optischen Eigenschaften wie SLR, PGR, chiro optische Effekte, Photo Stromverstärkung durch Heiße Elektronen Injektion und vieles mehr. Diese kollektiven Effekte bieten ein großes Potenzial für verschiedene Anwendungen, z. Plasmon-Lasern, hochwertige Plasmon-Sensorik, oberflächenverstärkte Raman-Streuung (SERS), optische Metamaterialien usw.[127,207,208]

In meiner Diplomarbeit konzentrierte ich mich hauptsächlich auf die TASA von sphärischen Goldnanopartikeln auf PDMS-Matrizen in 2D-Form. Wir haben uns auf zwei unterschiedlich beschichtete Partikel Typen Au_BSA-System und Au_PEG-System konzentriert. Schließlich erreichten wir eine SLR-Reaktion in diesem plasmonischen 2D-Gitter, die aufgrund der Flexibilität der Schablone in Echtzeit abstimmbar war. Wir diskutierten den Aufbau geordneter Nanopartikel-Arrays, die über 100 Verformungs Zyklen eine enge und mechanisch einstellbare Spiegelreflexkamera zeigen. Die Anordnung beruht auf der gerichteten Selbstorganisation kolloidaler Nanopartikel in durch weiche Lithographie gebildeten Matrizen. Wir haben die Anordnung mit breiter / enger Partikelgrößenverteilung gezeigt und ihren Einfluss auf die optische Qualität diskutiert. Die mechanische Echtzeit Abstimmung und Zugänglichkeit nicht entarteter SLR-Modi birgt ein großes Potenzial für plasmonisches Lasern[110], Biosensoren[209], oder kolorimetrische Sensoren.[210] Wir zeigen, dass sich die optischen Eigenschaften des Systems beim Anlegen von Dehnungen dramatisch ändern, wenn sich das Array von ändert ein quadratisches Gitter zu einem rechteckigen Gitter. Wichtig ist, dass die optische Qualität beim Dehnen des Systems nicht verloren geht und die SLR-Peaks über einen Wellenlängenbereich von 70 nm verschoben werden können. Darüber hinaus eröffnet das einachsig gespannte Gitter die Möglichkeit, eine optische Anisotropie einzuführen **(Kapitel 2.1).**

Weiter in dieser Richtung etablierte ich TASA von BSA-beschichteten AUNPs in 1D-Gittergeometrie. Insbesondere wurde die Montage direkt auf einem Fotolackgitter durchgeführt, was die Anzahl der an der Herstellung beteiligten Schritte verringert und eine einfache und schnelle Herstellung ermöglicht. Wir haben diesen Prozess auf TiO2- (200 nm dicker Film) und Goldfilm (35 nm) etabliert, die von einem Glassubstrat getragen wurden. Dies führt zu hybridisierten geführten Moden wie PGR und Resonanzen außerhalb der Ebene. Die hergestellte Struktur mit einer geeigneten Wahl der Entwurfsparameter unterstützt die Kopplung der plasmonischen Strahlungsmoden des Gold-NP-Gitters an die photonischen Moden des TiR2-basierten GMR bei normalem Einfall; Dies führt zur Bildung hybridisierter Zustände, die durch Variation des AOI eingestellt werden können. Die hybridisierte Struktur wird im Vergleich zu ihren konstituierenden Resonanzgeometrien optisch charakterisiert, zusammen mit einer eingehenden Untersuchung dieser Resonanzen durch numerische Simulationen, um die Kopplung zweier Resonanzen zu verstehen. Auf der Grundlage der Diagramme des elektrischen Feldes wird auch ein Matrixübertragungsmodell erstellt, das allgemein auf ähnliche Fälle der Kopplung verschiedener Resonanzen angewendet werden kann. Mit einem System, das auf dem Gebiet der modalen starken Kopplung völlig neu ist, haben wir seine mögliche Anwendung in der RI-Erfassung erkannt, bei der die Empfindlichkeit eines GMR-Geräts durch Mischen von plasmonischen Signaturen ohne großen Auflösungsverlust weiter verbessert werden kann **(Kapitel 2.2)**.[211]

In unserer oben erwähnten Studie haben wir bewiesen, dass Wellenleiter-Plasmon-Polariton in den Nanopartikelketten mit TE-polarisierter Anregung angeregt werden kann. Weiterhin möchten wir diese Erkenntnis hinsichtlich der Photostromerzeugung mit den kolloidalen Bausteinen erweitern. Eine spätere Pump- und Sondenmethode wird verwendet, um das Kopplungsregime in beiden Fällen zu verstehen: Metallstäbe mit TM und Metallnanopartikelketten mit TE-Polarisation. Ultraschnelle Spektroskopie könnte zeigen, dass sich das Elektron im Halbleitermaterial befindet. Wir haben ein ähnliches GMR-Gerät mit Nanopartikelketten entwickelt, bei dem die templatgestützte Selbstorganisation als Werkzeug zum Füllen der über LIL in großem Maßstab erzeugten Fotolackgitterkanäle verwendet wurde. Daher präsentieren wir eine Alternative zu den durch Elektronenstrahlen hergestellten Strukturen zusammen mit der Änderung der Polarisation mit der Einfachheit des kolloidalen Weges zur kostengünstigen und plasmonunterstützten Photostromverbesserung in großem Maßstab. Wir hoffen, dass dies die Roadmap für bessere Photokatalyse-, Photodetektions- und Sensoraktivitäten ebnen könnte **(Kapitel 2.3)**.[212]

Weiterhin haben wir das Stapeln von zwei Substraten durchgeführt, die Partikelkettenarrays umfassen. Dies ist eine einfache, aber leistungsstarke Strategie zur reversiblen Steuerung aller drei CD-Eigenschaften (Größe, Vorzeichen und spektrale Position). Der kolloidale Ansatz profitiert von einer kostengünstigen Herstellung und erzeugt gleichzeitig eine ausgeprägte chiroptische Reaktion. Die ausgedehnten superchiralen Felder, die in der Nähe der chiralen Nanostruktur entstehen, können die Empfindlichkeit bei der enantioselektiven molekularen Detektion erhöhen. Zu diesem Zweck können Analyten leicht in den Hohlraum des Partikelstapels eingebracht werden. Darüber hinaus erscheinen die stärksten CD-Modi im ersten (650-950 nm) und zweiten (1000-1700 nm) Transparenzfenster für biologisches Gewebe.[31] Folglich ist das eingeführte System gut zur Analyse biologischer Proben geeignet. Die Echtzeit-Abstimmbarkeit von CD durch erneutes Stapeln oder Komprimieren fördert die Entwicklung kompakter spektroskopischer Geräte und Lichtmodulatoren. Als einzigartiges

4. Zusammenfassung

Merkmal ermöglicht das eingeführte System eine spannungsinduzierte lokale Abstimmbarkeit der chiro-plasmonischen Eigenschaften. Das mechanistische Verständnis dieses Phänomens mittels mechanischer und elektromagnetischer Simulationen liefert die Grundlage für die Verallgemeinerung des beobachteten Effekts und die rationale Gestaltung chiroptischer Eigenschaften. Insbesondere dient dieser Mechanismus als neues Paradigma für das Design aktiver chiraler Geräte, hochintegrierter Sensoren und komplexer Strukturierung von Licht **(Kapitel 2.4).**[213,214]

Darüber hinaus ermöglicht das in diesem Zusammenhang entwickelte Amalgam von Top-Down- und Bottom-Up-Ansätzen den Zusammenbau von Nanopartikeln mit unterschiedlichen Geometrien und Zusammensetzungen (z. B. Kern-Schale-Partikel, Würfel, Quader, stabförmige Partikel von Au, Ag). Schließlich ermöglicht das etablierte rationale Design-Framework auch die Erweiterung des Ansatzes auf die suprakolloidale Struktur (Cluster, Hetero-Cluster usw.).[99,213,215,216]

6. Appendix

The information and figures in the Appendix are published in the Supporting Information of the underlying publications and are adapted within the respective permissions.

6.1.   laser interference lithography

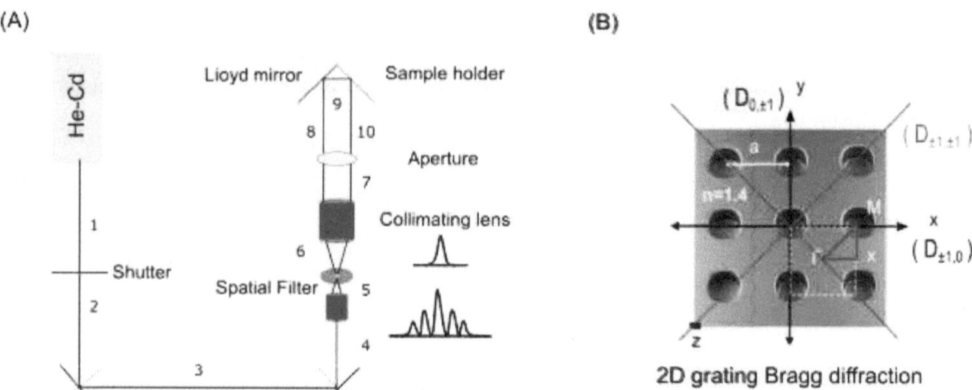

Figure 22: (A) Optical set-up for laser interference lithography via Lloyd's mirror technique. (B) Diffraction modes of a 2D square grating (nanohole array) with refractive index n and lattice constant a where Γ, X, M represent the symmetry points in the reciprocal space.

Having the advantage of large-scale fabrication with nanometer precision at low cost we opt for laser interference lithography where Lloyd's mirror interferometer technique was used to write the structures on a photoresist coated silicon substrate. In terms of optics, a laser beam ($\lambda = 325$ nm) follows a path as shown in (Figure 22A). After collimation, a single coherent beam converted into two coherent sources by the Lloyd mirror. The wave-fronts from these sources interfere at a plane where intensity pattern is given by $I(x) = 2A \{ Cos[(4\pi x/\lambda)\sin(\theta)] +1\}$ where A is the amplitude, λ is the wavelength of the laser beam and θ is the interference angle between two beams.[1] These intensity patterns, i.e. 1D line patterns, were recorded on a UV light-sensitive material. Depending upon the material type (positive/negative), the half angle between the beams (θ), sample rotation and energy dosage requirement (threshold), various kinds of patterns with different duty cycle and periodicity can be produced. In our experiments, we have used a positive photoresist in combination with BARC (back anti-reflection coating) for high resolution mask-less pattern printing. Figure 22B represents the Bragg diffraction order for a 2D grating.

6.2.   Soft molding

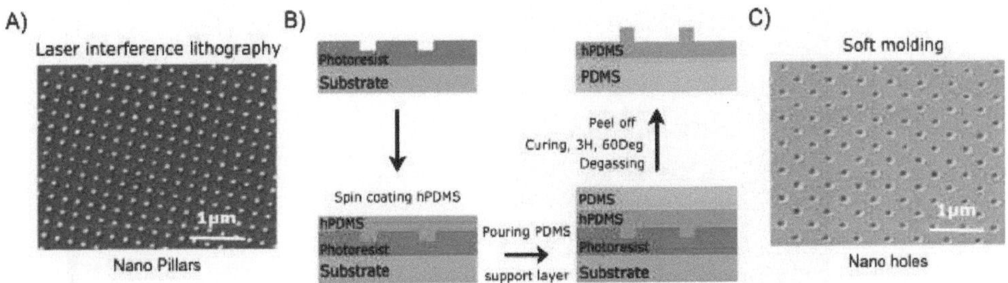

Figure 23: (A) SEM image of nanopillar array fabricated with LIL on a photoresist material with periodicity 400 nm. (B) Steps of soft molding procedure. (C) Nano holes array with periodicity 400 nm produced via replication of nanopillars array with soft molding technique.

Soft molding is considered as an unconventional method of writing since it is very much clear from the previous details of conventional methods that they have relatively high cost and low throughput techniques. These conventional methods such as e-beam are mainly restricted to planar fabrication and are incompatible with many problems such as three-dimension (3D) fabrication and scalability. These methods also use corrosive etchants, high energy radiation and relatively high temperature. This is also true that these techniques are highly energy consumptive and researchers are looking forward to alternatives (unconventional) techniques. Conventional methods are also limited towards selection of materials when it comes to patterning rather fragile materials and especially biological materials these techniques don't support that. Hence it is clear that regardless conventional or unconventional- any technique which has capability to prototype nanoscale features rapidly and inexpensively will be acceptable in future. There is huge demand in the direction of real-to-reel, parallel, scalable, cost-effective and versatile methods.

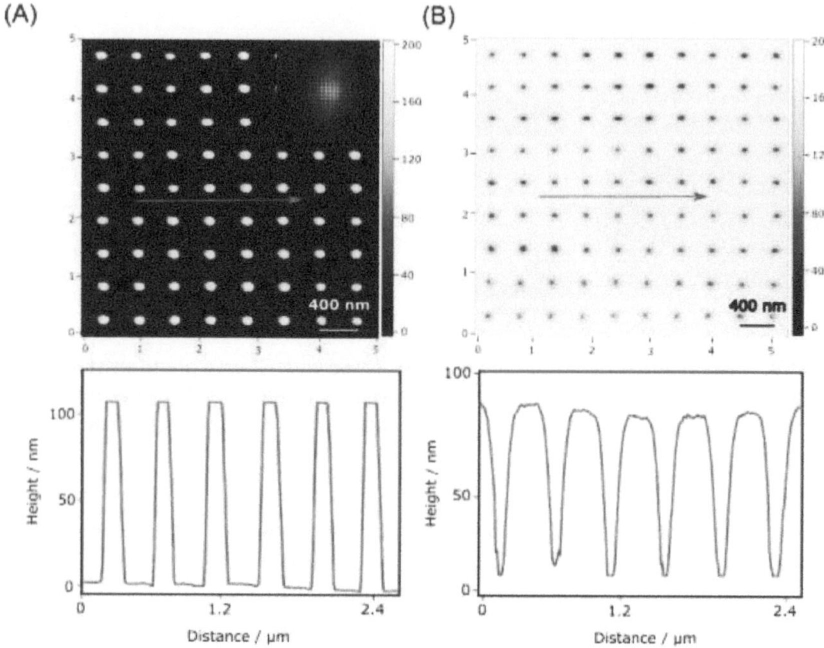

Figure24: (A) AFM height profile of nanopillars patterned by LIL. (B) AFM height (depth) profile of nanoholes replicated via soft-molding. Inset: Fast Fourier transform of the images.

6.3. Determine fill factor of plasmonic lattice

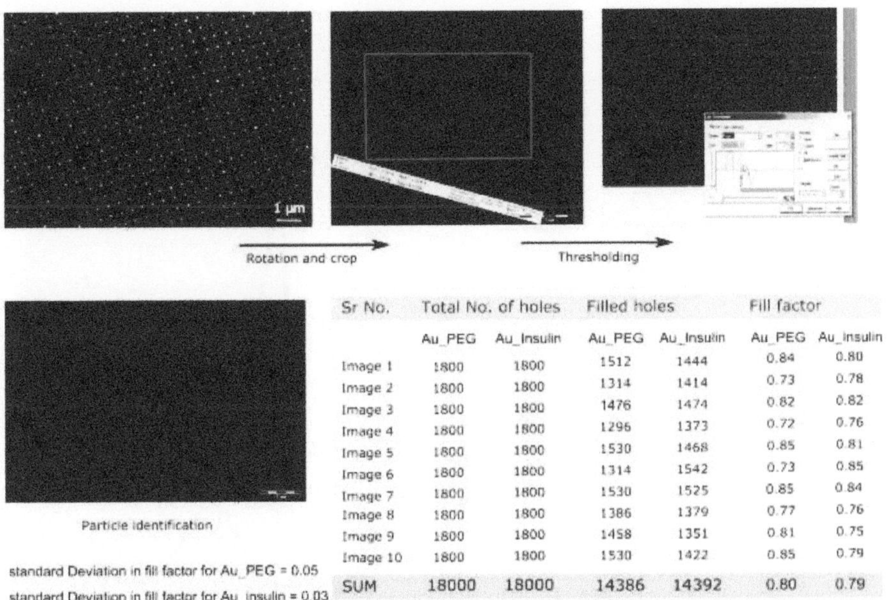

Sr No.	Total No. of holes		Filled holes		Fill factor	
	Au_PEG	Au_Insulin	Au_PEG	Au_Insulin	Au_PEG	Au_Insulin
Image 1	1800	1800	1512	1444	0.84	0.80
Image 2	1800	1800	1314	1414	0.73	0.78
Image 3	1800	1800	1476	1474	0.82	0.82
Image 4	1800	1800	1296	1373	0.72	0.76
Image 5	1800	1800	1530	1468	0.85	0.81
Image 6	1800	1800	1314	1542	0.73	0.85
Image 7	1800	1800	1530	1525	0.85	0.84
Image 8	1800	1800	1386	1379	0.77	0.76
Image 9	1800	1800	1458	1351	0.81	0.75
Image 10	1800	1800	1530	1422	0.85	0.79
SUM	18000	18000	14386	14392	0.80	0.79

Figure 25: Fill factor calculations were performed on Scandium software where 10 SEM images (ESB detector) were used to identify the particles on 150 x 150 µm^2 area. Table shows the quantitative analysis of the images investigated.

We have analyzed 10 SEM images (ESB detector) each 20 µm x 20 µm for fill factor calculation. The process follows from rotating and cropping the SEM images to thresholding and particle identification. Fill factor was denoted by the ratio of filled holes to the total number of holes. Average of 10 images was considered as the overall fill factor over 150 x 150 µm2 areas.

6.4.   2D plasmonic lattice of Au_BSA under strain

(A)

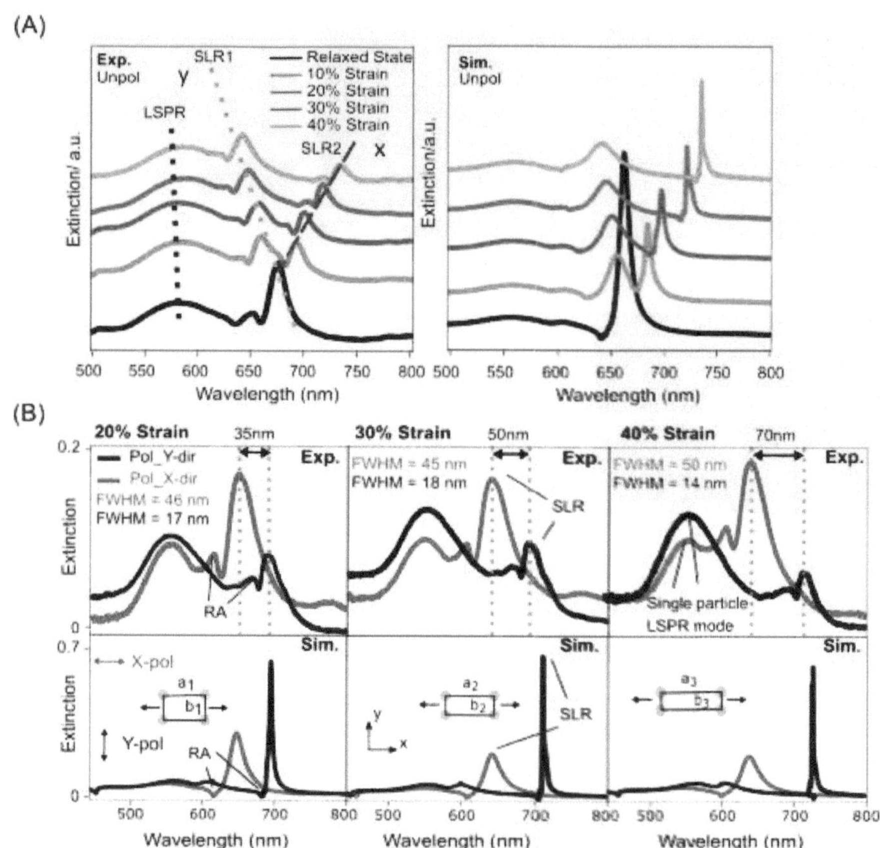

Figure 26: (A) Experimental and simulated spectra of gold nanoparticle lattice (lattice constant = 440 nm) under various strains. (B) Polarization dependent switching of the lattice resonance modes experimental (B top row) and simulated (B bottom row).

Increasing the strain gradually from 0% to 40% transforms the square lattice into a rectangular one with lattice constant a (along applied strain direction) and b (along perpendicular direction) (Figure 26A). Two different lattice constants in x- and y-direction result in two different SLR modes (both are band-edge in nature). Due to the stretching-induced symmetry-breaking one of the SLR shifts towards higher wavelengths (violet colour marker in Figure 26A), whereas the other one is blue shifted (orange colour marker in (Figure 26A) for unpolarized light illumination. The spectral shifts are in good agreement with finite-difference time-domain (FDTD) simulations for strained lattices. The lattice constant in y-direction was calculated from the applied strain in x-direction based on the tabulated Poisson's ratio for

PDMS. The two different SLR modes in rectangular lattices can easily be switched from one another when using linearly.

polarized excitation, as depicted in Figure 26B for different strains. Note, that the SLR induced by polarization in x-direction is at lower wavelengths as compared to y excitation. At first glance, this is counterintuitive as the periodicity in x-direction is larger. However, this can be understood by considering every plasmonic particle as a dipole antenna. Excitation along the x-axis induces electron oscillations in parallel orientation for each particle.

6.5. Characterizing order inside a 2D lattice

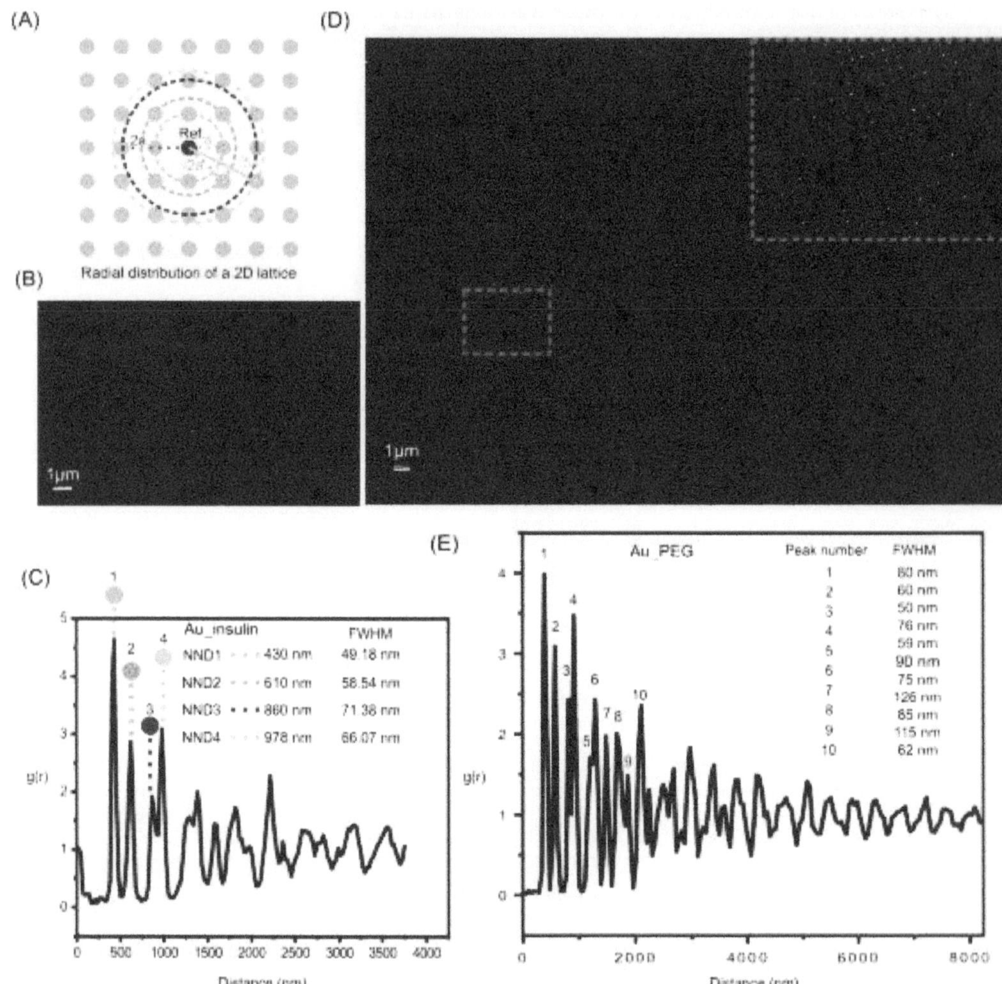

Figure 27: (A) Schematic representation of the radial distribution in a 2D square lattice with lattice constant a. (B) SEM image (ESB detector) under consideration for evaluation. Definition of abbreviations: Nearest neighbor distance (NND). (C) Radial distribution function g(r) of a 2D assembled lattice of insulin coated particles (diameter 83 nm) with lattice periodicity 440 nm. (D) Large area (38 x 26 µm2) SEM image of the PEG-coated gold nanoparticle (77 nm in diameter) lattice (inset shows highlighted area in red color for better contrast). (E) Radial distribution in large scale lattice with peak positions and FWHM.

Peaks in the radial distribution graph provide information about the nearest neighbor distances. In case of an ideal 2D square lattice the first four distances are a, √2a, 2a, √5a. We have calculated similar distances for the given SEM image via peak position identification. Values are in good agreement (±10 nm) for a lattice of 440 nm periodicity.[217] Similar radial distribution was also determined for large area PEG-coated particle lattice of 38 x 26 μm^2 domain size. Overall order was maintained up to 3 μm.

6.6.   Template-assisted colloidal self-assembly

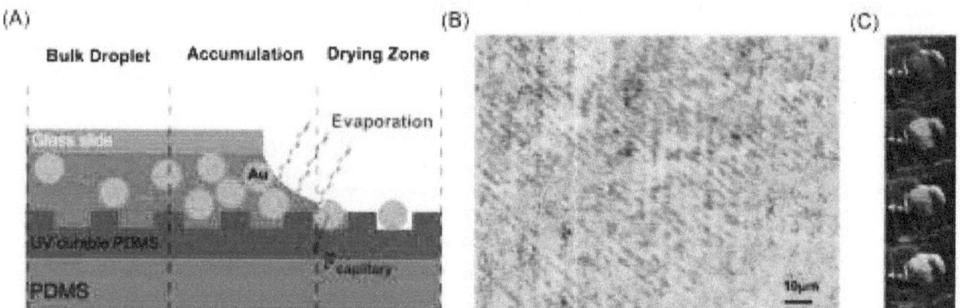

Figure 28: (A) Mechanism for the template-assisted capillary self-assembly. (B) Dark field image of a plasmonic lattice with lattice constant 440 nm. (C) Large scale 2D plasmonic grating photographs captured at different angles of incidences.

With the motorized movement of the template (1 μm s^{-1}), the liquid meniscus recedes across the recessed holes. The evaporation of liquid facilitated close to the three-phase contact line drives a convective flow of particles towards the meniscus and consequently accumulates them. All experiments were carried out with substrate temperature of 12 K above the dew point to provide a constant evaporation rate and compensate for changes in room temperature and relative humidity. The local concentration reduces the Brownian motion of nanoparticles and thereby increases the probability of particles residing in a feature during capillary break-up, which is a prerequisite to obtain high filling rates. This also means that the particles have to stay colloidally stable even at high concentrations and the formation of irreversible aggregates has to be avoided. In our system, different coatings (PEG: mainly steric; insulin: electrostatic repulsion, zeta potential ζ ~(-30 mV) at pH 92) ensured sufficient inter-particle electrostatic repulsion.

Furthermore, the net repulsive particle-substrate interaction allowed effective dragging of the particle front, so that no particles were randomly adsorbed at the top surface but only were placed deterministically inside the holes. Precise tuning of contact angle and surface tension using a mixture of anionic (sodium dodecyl sulphate, SDS) and nonionic (Triton X-45) surfactants[76,107] balanced the components of the capillary force at the meniscus acting parallel and normal to the substrate. That way the gold nanoparticles are effectively dragged along and are only deposited where they feel a sidewall of the topographical features counteracting this motion. The combination of 0.25 mM SDS and 0.025 wt % Triton X-45 yields ~27 mN m^{-1} surface tension,[196] which was demonstrated to be a prerequisite for successful capillary nanoparticle assembly.[107] Note that the dense accumulation zone forming at the meniscus ensures sufficient supply of particles to provide high filling rates for all array periodicities (380-440 nm) without drastic change in assembly parameters. Dark field microscopy supports the good coverage over a large area as a predominant green color originates from the single 83 nm gold particle scattering.6 Due to imperfections in the depth of the LIL-produced master, sometimes one-hole hosts two particles, which explains occasional red color dots in the image (Figure 28B). Diffraction of white light over the whole substrate (Figure28C) with different angles of incidence also shows the quality of periodic arrangement in the system.

(A)   (C)

(B)

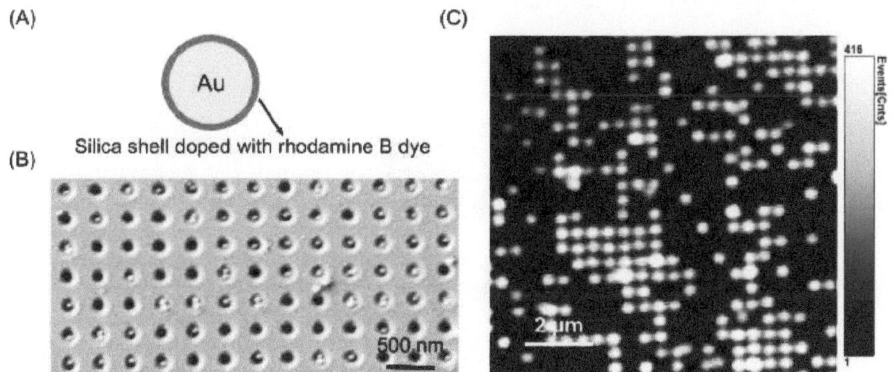

Figure29: (A) Gold nanoparticle (80 nm in diameter) coated with rhodamine B dye doped silica shell. (B) Assembly of such particles in a 2D grating of PDMS fabricated with combination of LIL and soft molding. (C) Fluorescent image of such a lattice.

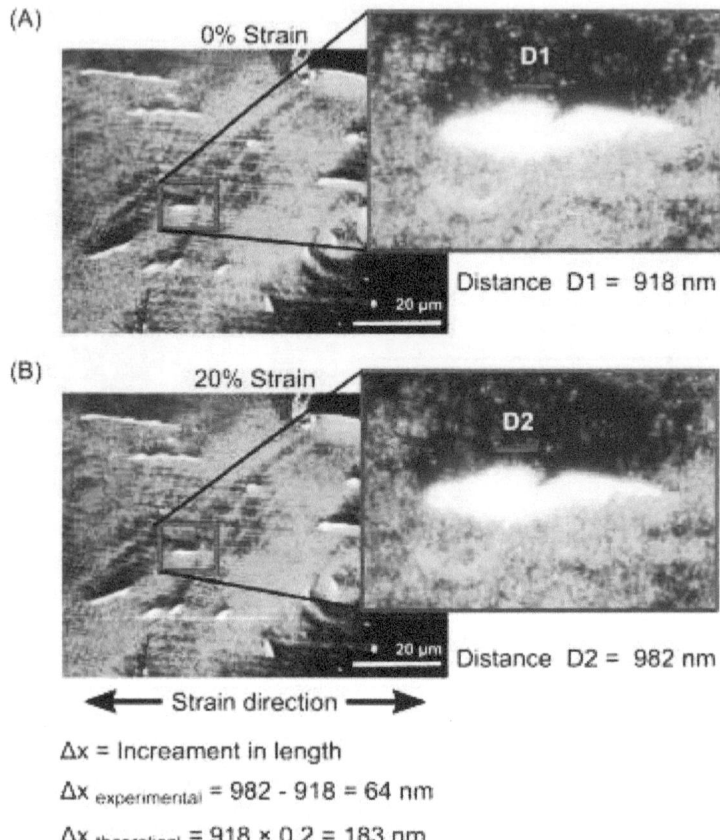

Figure 30: (A&B) Dark field image of a plasmonic lattice (lattice constant = 440 nm) under 0 and 20 strain.

A plasmonic lattice with lattice constant 440 nm is considered for finding out the particle movement with index matching layer on top. A dust particle was considered as a marker and distance between two nanoparticles was calculated. D1(918 nm) is the distance in a relax state and D2 (982 nm) is the distance for a 20% strain case. Theoretically D2 should become 183 nm more than D1 but experimentally only 35% force transfer was observed which is also supported by the SLR measurement (37% force transfer experimentally).

6.7.   Out of plane lattice resonance in 1D and 2D lattices

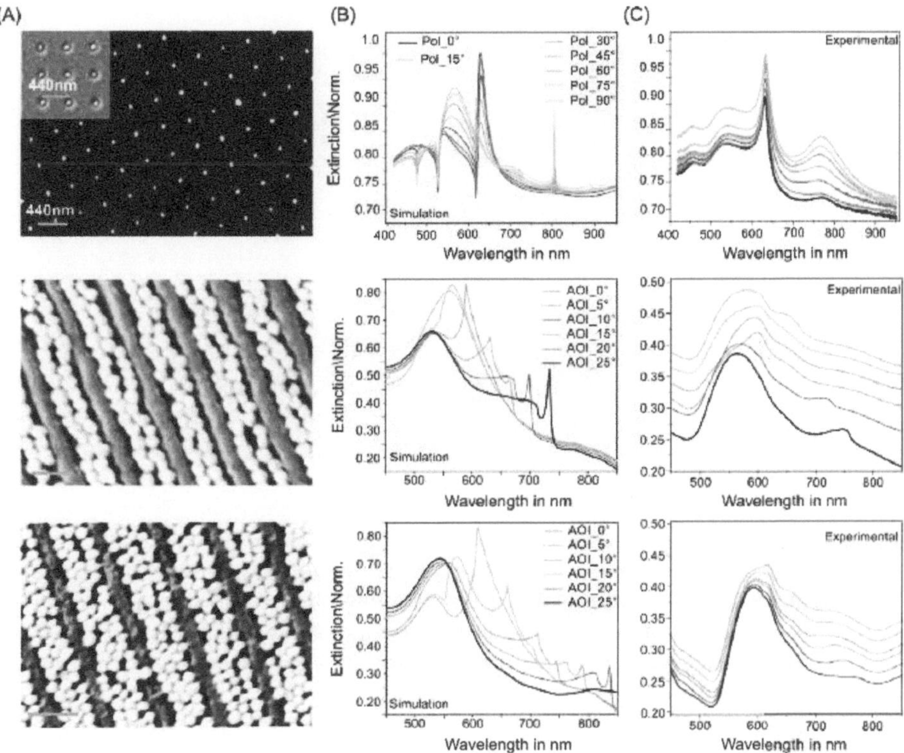

Figure 31: (A) SEM images of 2D plasmonic lattice of gold nanoparticle and 1D lattice of gold nanoparticle with dimer and trimer unit cell respectively. The scale bar in dimer case is 330 nm and for trimer case 400 nm. (B) Simulated optical spectra of a 2D lattice for various polarization, 1D lattice (with different periodicity) for various angles of incidence. Measured extinction spectra of 2D lattice for various polarization and for 1D lattice of different periodicity for given angle of incidence.

6.8. E-Field distribution at out of plane SLR mode for 1D lattices of various periodicity with AOI 20°

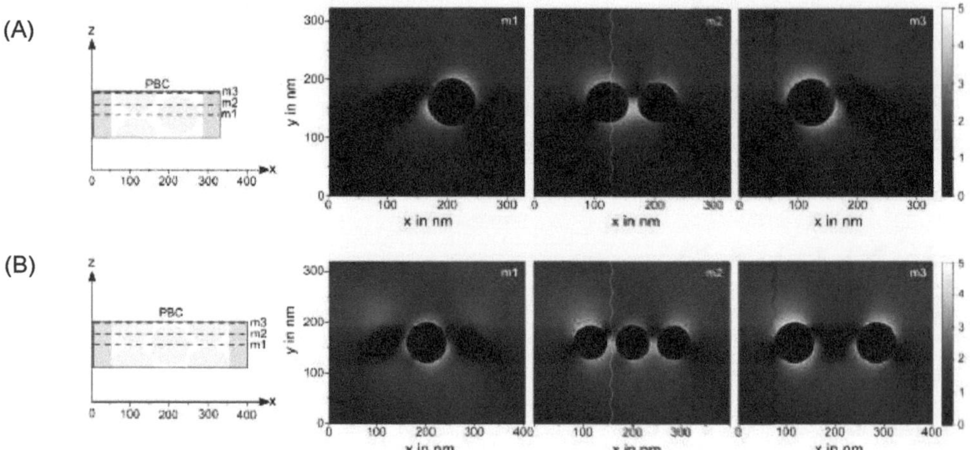

Figure 32: (A) simulated normalized E-field intensity plot at the different location of dimer 1D gold nanoparticle lattice. The spectra is plotted for a particular angle of incidence (AOI_20°) with TM excitation at the out of plane lattice resonance wavelength. **(B) Simulated normalized E-field intensity plots at the various locations of the trimer 1D lattice of gold nanoparticle where locations are shown with m1, m2, m3. The spectra is plotted for AOI_20° with TM excitation at the out of plane lattice resonance wavelength.

6.9. Refractive index of PDMS and UV-PDMS

(A)   (B)

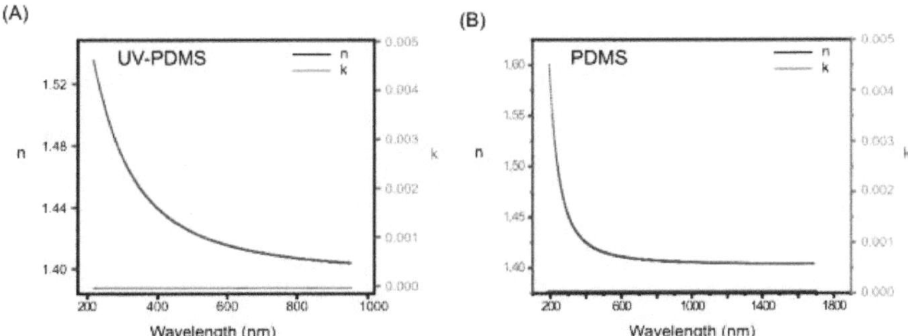

Figure 33: (A) and (B) represent wavelength dependent optical constants of UV-PDMS and PDMS Sylgard (mixing ratio, 10:1) measured by spectroscopic ellipsometry respectively.

6.10. Refractive index measurement for sensing

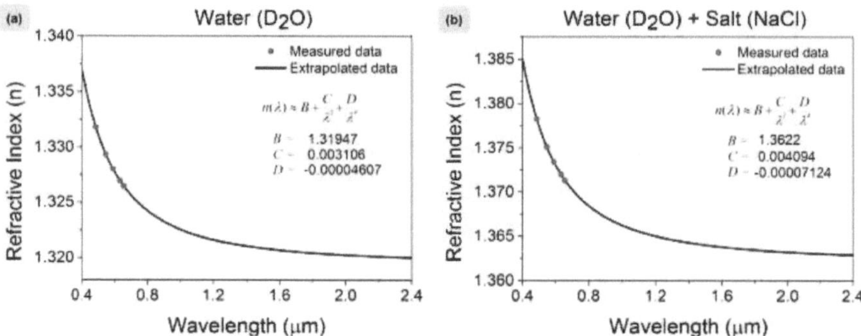

Figure 34: Refractometer data: The real parts of refractive index (r. i.) are plotted for (a) heavy water (D2O) and (b) salt solution (NaCl in D2O, 0.3g/ml) using known values measured at five different wavelengths by a refractometer.

The refractive indices of pure water and salt solution as described in the main article are measured using a digital refractometer for five discrete wavelengths. Using Cauchy's dispersion equation, the coefficients B, C, and D are determined which are further used to extrapolate values of a real part of the refractive index (n) for the entire range of wavelength scan from 400-2400 nm. Figure 34 (a) and (b) show the 'n' values of pure water and salt solution respectively as a function of wavelength. These data are also incorporated within the simulation model to observe the sensitivity of different resonant structures on changing surrounding media from pure water to salt solution.

6.11. Optical constants of TiO2, ma-N 405 photoresist and glass substrate measured from spectroscopic ellipsometry

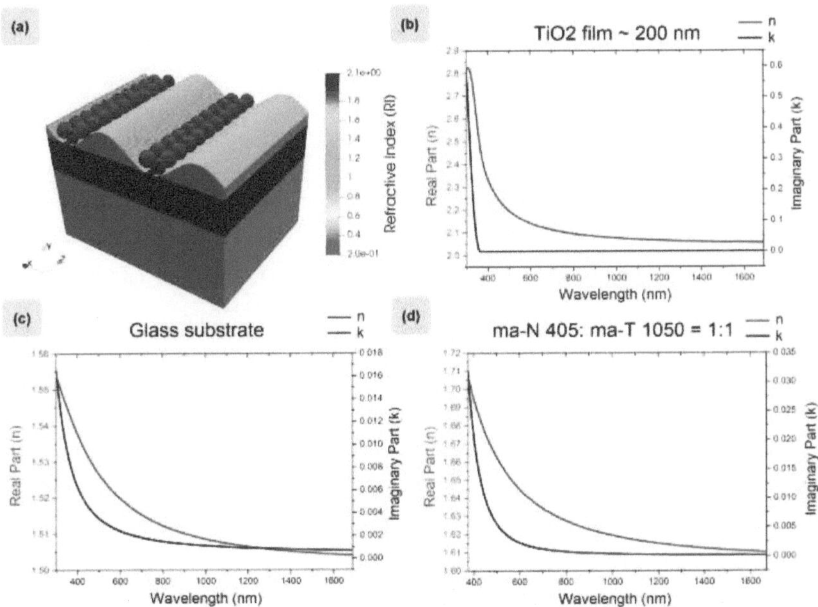

Figure 35: Measured refractive index data used for FDTD simulations: (a) 3D index data of the hybrid structure observed at λ = 851 nm. (b), (c) and (d) Real and imaginary values of the refractive index data of TiO_2, glass and diluted photoresist respectively measured through spectroscopic ellipsometer and included in the simulation model.

Figure 35a shows the refractive index profile of the hybrid structure at 851 nm with optical constant data fitted from experimentally values using spectroscopic ellipsometry (RC2-DI, J.A. Woollam Co., Inc.). Thin films of TiO_2 and diluted photoresist along with bare glass substrate are separately measured for the entire range of operation (400-2400 nm) and plotted in Figure 35 (b-d). These data are incorporated in Lumerical software to provide materialist properties to the designed models that are used throughout the article for comparative analysis of different resonance mechanisms.